Agricultural and Food Systems Sustainability: The Complex Challenge of Losses and Waste

Editors

Alessandro Suardi
Nadia Palmieri

MDPI • Basel • Beijing • Wuhan • Barcelona • Belgrade • Manchester • Tokyo • Cluj • Tianjin

Editors

Alessandro Suardi
CREA Research Centre for Engineering and Agro-Food Processing
Italy

Nadia Palmieri
CREA Research Centre for Engineering and Agro-Food Processing
Italy

Editorial Office
MDPI
St. Alban-Anlage 66
4052 Basel, Switzerland

This is a reprint of articles from the Special Issue published online in the open access journal *Sustainability* (ISSN 2071-1050) (available at: https://www.mdpi.com/journal/sustainability/special_issues/agricultural_food_systems_sustainability).

For citation purposes, cite each article independently as indicated on the article page online and as indicated below:

LastName, A.A.; LastName, B.B.; LastName, C.C. Article Title. *Journal Name* **Year**, *Volume Number*, Page Range.

ISBN 978-3-0365-6013-7 (Hbk)
ISBN 978-3-0365-6014-4 (PDF)

Agricultural and Food Systems Sustainability: The Complex Challenge of Losses and Waste

Contents

About the Editors

Alessandro Suardi

Dr. Alessandro Suardi is a senior researcher and agricultural engineer for forestry and the environmental management with a PhD in science and technology. A senior researcher at the Council for Agricultural Research and Economics (CREA) in Italy since 2008, he has worked on nine national and eight international projects for the development of new prototypes to optimize agricultural mechanization logistics, improve farm mechanization, harvesting and management, valorise agricultural residues, and perform environmental sustainability assessments of agricultural supply chains with the Life Cycle Assessment method (LCA). He is the author of more than 100 national and international scientific publications.

Nadia Palmieri

Dr. Nadia Palmieri is researcher and economist, with both a Ph.D and a national scientific qualification in agricultural economics. She is researcher at the Council for Agricultural Research and Economics (CREA) in Italy. Her main areas of research are consumer behaviour, willingness to consume, willingness to accept, agricultural economics as well as environmental and economic assessment of agricultural systems. On these subjects, she was the scientifist responsible for the economic aspect of 8 research projects and has published more than 50 papers.

sustainability

MDPI

Article

Comparison between Two Strategies for the Collection of Wheat Residue after Mechanical Harvesting: Performance and Cost Analysis

Alessandro Suardi [1], Walter Stefanoni [1,*], Simone Bergonzoli [2], Francesco Latterini [1], Nils Jonsson [3] and Luigi Pari [1]

1 Consiglio per la Ricerca in Agricoltura e l'analisi dell'Economia Agraria (CREA)-Centro di Ricerca Ingegneria e Trasformazioni Agroalimentari, Via della Pascolare, 16, 00015 Monterotondo (RM), Italy; alessandro.suardi@crea.gov.it (A.S.); francesco.latterini@crea.gov.it (F.L.); luigi.pari@crea.gov.it (L.P.)

2 Consiglio per la Ricerca in Agricoltura e l'analisi dell'Economia Agraria (CREA)-Centro di Ricerca Ingegneria e Trasformazioni Agroalimentari, Via Milano, 43, 24047 Treviglio (BG), Italy; simone.bergonzoli@crea.gov.it

3 Research Institutes of Sweden (RISE), Ultunaallén 4, 756 51 Uppsala, Sweden; nils.jonsson@ri.se

* Correspondence: walter.stefanoni@crea.gov.it; Tel.: +39-06-9067-5205

Received: 25 May 2020; Accepted: 12 June 2020; Published: 17 June 2020

Abstract: The growing population worldwide will create the demand for higher cereal production, in order to meet the food need of both humans and animals in the future. Consequently, the quantity of crop by-products produced by cereal cropping will increase accordingly, providing a good opportunity for fostering the development of the sustainable supply chain of renewable solid fuels and natural feedstock for animal farming. The conventional machineries used in wheat harvesting do not guarantee the possibility to collect the chaff as additional residue to the straw. The present study investigated the possibility to equip a conventional combine with a specific device, already available on the market, in order to collect the chaff either separately (onto a trailer), or together with the straw (baled). The total residual biomass increased by 0.84 $t{\cdot}ha^{-1}$ and 0.80 $t{\cdot}ha^{-1}$ respectively, without negatively affecting the performance of the combine when the chaff was discharged on the swath. Farmers can benefit economically from the extra biomass collected, although a proper sizing of the machine chain is fundamental to avoid by-product losses and lower revenue.

Keywords: biomass; bioenergy; straw; combine harvester; chaff; by-product

1. Introduction

1.1. Framework

The use of non-renewable sources for meeting the fast-growing energy demand worldwide could trigger negative effects on the environment in terms of pollution. On the other hand, as the worldwide population is expected to exceed 9 billion people by 2050 (FAO), the production of several key commodities will also increase accordingly, in order to meet the fast-growing demand for food. The production of cereals is expected to grow from the annual 2.1 billion tons up to 3 billion tons by 2050 if animal feeding is also included [1]. Consequently, the ongoing conflict on land use for food and non-food crops will be more serious if new strategies are not promptly undertaken. Regarding the bioenergy production, the European policy is keen to promote the utilization of agroforestry residues over the plantation of energy crops [2], by applying stringent regulations, in order to meet the climate and energy targets set in the EU 2030 framework [3,4]. Hence, a possible strategy could be improving the collection and the utilization of residual biomass that is normally produced in cereal cropping, but not effectively exploited yet [5]. During the harvesting of cereals, for example, in addition to the collection of grains, a large quantity of residual biomass is usually produced as straw and chaff.

Among them, straw has been exploited for a long time as natural bedding for animals [6] and, recently, as a valid source for energy production or as raw material for industrial processes. In terms of energy, one hectare of cereal straw is approximately equivalent to 200 L of oil [7] if considered as solid biofuel. However, the biochemical properties of ligno-cellulosic materials, as straw, make it suitable for further industrial processing. For instance, Fang and Shen [8] reported the suitability of straw for paper and paperboard production, Hýsek et al. [9] highlighted the possibility to exploit cereal residues for composite material production, while Swain et al. [10] investigated the hydrolyzation of cellulose and the hemicellulose of straw into fermentable sugars, which are particularly attractive for bioethanol production industries. Recently, it has been found that winter wheat straw can be returned to soil as biochar to enhance the yield in corn and peanut cultivation [11]. Conceptually, the development of a comprehensive, efficient and sustainable straw supply chain can bring benefits to many sectors and to developing countries as well [12].

On the other hand, the chaff, as the finest part of the grain residue, is normally lost on the ground after mechanical harvesting. In wheat crops, chaff is available at the rate of the 17% of the grain yield [13], and if considering the European annual production of wheat and spelt estimated in, approximately, 138 Mt [14], the whole European biomass supply chain could benefit from the collection of 23 $Mt{\cdot}yr^{-1}$ more of biomass. This would contribute to increase the availability of solid biofuel for the production of energy, particularly if baled with the straw [15]. Chaff palletization is also possible, but only if provided as loose product [16], as well as for the production of second generation bioethanol [17].

Nevertheless, the collection of chaff has already been investigated in agriculture, as a promising tool in organic farming of cereal grain for reducing the accumulation of weed seeds in the soil over time [18,19]. In Australia, different mechanical devices were invented and tested on field, with the specific purpose of removing the chaff for decreasing the amount of weed seeds [20–22]. The chaff was then arranged in small hips or in narrow strips for being burnt afterwards. The possibility to collect chaff for purposes different from weed seeds control, has already been investigated under the economic aspect by Unger and Glasner [23], whose study revealed that the exploitation of that kind of residue is feasible. Although the simultaneous harvesting of wheat grains and chaff has been recently investigated [19,24–26], the literature still lacks of specific data from in-field experiment.

Actually, mainly due to the lack of knowledge on the specific devices already available on the market, uncertainty on the harvesting system to adopt and due to the lack of a dedicated supply chain for an effective exploitation, the chaff still remains an untapped biomass. There is a real need to evaluate the cost effectiveness and the performance of systems for harvesting chaff in order to foster the utilization of this biomass, depending on the final use, and to stimulate the development of a dedicated value chain. According to this, the aim of the study consisted in filling this knowledge gap and providing a deeper understanding of the possibility to enhance the current wheat harvesting method, in order to improve the quantity of biomass collected, including the chaff.

1.2. Main Chaff Utilization

The chaff from cereal crops can be handled differently according to its final utilization. More recently, the chaff is thought as a source of biomass for energy use, but others are known in literature. For instance, in Australia, harvest weed seed control (HWSC) systems have been developed and tested for years, providing good results in terms of alternative strategy for weeds control. Walsh, Newman, and Powles in 2013 [20] reviewed the following systems: chaff chart, narrow windrow burning, bale direct and Harrington seed destructor (HSD). The first two of them accumulate the chaff, either in heaps, or in a narrow windrow (50–60 cm wide) on the field for direct burning. Among the other two systems, apart from the HSD that mechanically destroys the seed weeds, the direct baling strategy provides multiple choices for chaff utilization. In fact, the chaff is baled with the straw as soon as they exit the cleaning shoe of the combine harvester. Indeed, baling them addresses two main problems: the removal of weed seeds and the collection of biomass for livestock (both feeding [6,13] and natural bedding). The presence of chaff into straw bales also increases the adsorbent capacity of natural bedding [27]. Even poultry farming can benefit from loose chaff availability on the market. A direct interview with a local farmer in France highlighted the positive effects, noticed by farmers, on the welfare of the animals that could scratch around in search for broken kernels and weed seeds, which, in turn, contributed to overall feeding. The same experience was reported by Italian farmers. The use of similar cereal residues is reported in literature as a valid source for littering. Anisuzzaman and Chowdhury reported that rice husk was a good litter material for rearing broilers [28] and it also has a high adsorbent capacity if compared with sawdust [29]. Chaff could also be suitable for further processing, like briquetting, and used for multiple purposes. Akerlof [30] reported the possibility of producing briquettes of soybean chaff for meeting the needs of livestock in providing complete feeding, whereas spelt chaff has been proven to be a good raw material for the production of briquettes for non-feeding purpose, who exhibited different mechanical properties according to the temperature of compression applied [31]. Wheat chaff applications are not fully studied in the sense of both feeding or not-feeding purposes. The unviability of specific mechanical machines able to collect it without increase in the harvesting costs, has probably limited the research in that direction. For this reason, this study addresses an important issue for the development of new production chains based on cereal residues, showing two possible chaff collection logistics, the limits and operating costs of the technologies used, laying the foundations for the development of possible supply chains that are currently underdeveloped or, in some cases, non-existent. In the framework of the H2020 AGROinLOG project [32], a specific test in the Halland region (Sweden) was carried out, to provide evidences on the possibility of improving the conventional supply chains in wheat harvesting, for increasing the overall residual biomass collectable in the field. Specifically, the aim of the test was to evaluate if it is possible to accomplish such a task by equipping a conventional harvester combine with a dedicated device for chaff recovery, already available on the market and manufactured by the Thierart firm (Thierart, Le Châtelet-sur-Retourne, France) [33]. The device permitted one to flow the chaff, either onto a towed trailer, or on the straw swath produced by the combine harvester. Therefore, both chaff collection methods were tested: loose in a towed trailer (CoT), or baled together with the straw (CoS). The trailer was connected to the combine harvester, therefore no tractor was required for towing it. The amount of biomass potentially collectable as grains, straw and chaff was quantified, as well as the performance and quality of the work of all machines involved in the two supply chains. The loss of seeds, straw and chaff were recorded and an evaluation of the harvesting operating costs was carried out.

2. Materials and Methods

2.1. Field Site and Experimental Design

The test was performed at Lilla Bösld (Halland region, Sweden) (56°35′48″ N 12°57′33″ E) in the 35th week of 2019 (Figure 1). The field, 15 m a.s.l., exhibited a negligible value of slope.

The wheat (*Triticum* spp.) variety "Julius" was sown in medium clay soil type (24–29% of clay) in September 2018, with a seeding rate of 220 kg·ha^{-1} and cultivated in conventional farming. Fertilization was carried out with 150 kg·ha^{-1} of PK 11–21 and 500 kg·ha^{-1} of Nitrogen fertilizer (27% N + 9% SO3) and 200 kg·ha^{-1} of calcium nitrate. For the weed control 1 L·ha^{-1} of MCPA and 15 g·ha^{-1} of Express 50 (wetting agent 0.1 L·ha^{-1}) were used. For the fungus control, 0.5 L·ha^{-1} of Ascra Xpro was applied.

Within the field, a homogeneous area of 3 ha was preliminarily selected. The surrounding wheat was harvested and the whole biomass removed, in order to avoid edge effects and biased measurement. The selected area was then divided into three blocks, each of them sub-dived in two rectangular shaped plots measuring approximately 0.5 ha. Thus, three random replications per treatment were obtained, for a total of six plots. The chaff was collected in two different ways (treatments): either discharged on the swath (CoS) or collected on a trailer (CoT).

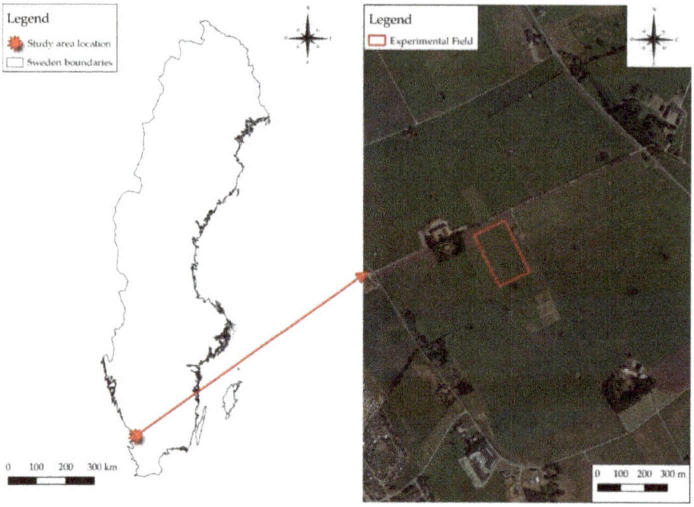

Figure 1. Map of the experimental field in Halland region of Sweden.

2.2. *Pre-Harvest Tests: Theoretical Biomass Assessment*

For management reasons, the test was split into two consecutive days: the first day was dedicated to the pre-harvesting activities and combine harvesting; the following day occurred the baling operation and post-harvesting activities. Before harvesting, the whole plants of 10 samples areas of 1 m^2 randomly chosen were hand harvested. Culms and spikes were weighed separately. Successively, all spikes and a representative sample of culms were put in sealed bags and shipped to the laboratory of Research Centre for Engineering and Agro-Food Processing (CREA) for further measurements as: theoretical yield of grain and chaff, dry weight and moisture content.

In the laboratory, by using a stationary thresher (PLOT 2375 Thresher, Cicoria Company, San Gervasio, Italy), kernels were separated from the rest of the spikes (rachis, lemma, glumes and palea). The dry weight and moisture content of culms, kernels and chaff was assessed according to the EN ISO 18134-2:2017 [34] standard.

2.3. *Equipment*

The contractor provided all the machines required for the test. Settings of the combine harvester, as well as the baler, were maintained at a constant rate throughout the experiment.

2.3.1. Combine Harvester and Recovery System

A combine harvester New Holland TX68 with a conventional threshing drum, straw walker and cleaning shoe was used to perform the test. The header was 7.27 m width and it was specifically designed for cereal harvesting. The machine was driven by a 209 kW diesel engine and the chassis was comprehensive of a dedicated hitch for trailer towing.

The device for the chaff recovery was installed at the end of the cleaning shoe of the combine harvester. As shown in Figure 2, the device is made of a tank that receives the chaff from the cleaning shoe; within it, there is a steal-made screw that delivers the chaff to the two-stage turbine which, in turn, blows it through the outlet. According to the company Thiérart [33], the device requires a minimum of 45 $L \cdot min^{-1}$ of hydraulic oil flow rate to work properly and the cutting bar of the combine harvester should not exceed 5.5 m in width to properly manage the chaff flow.

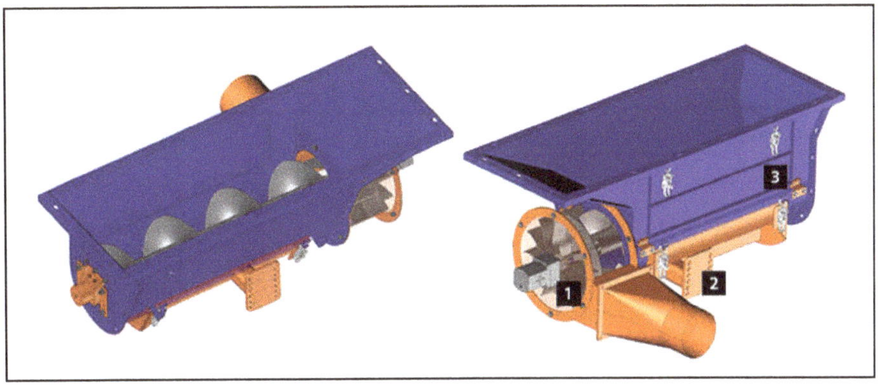

Figure 2. Device developed and patented by Thierart for chaff recovery: (1) two stages turbine; (2) specific support for the installation; (3) access hatch to the screw for inspection (source: https://www.menuepaille.fr/materiels/turbine-a-double-etage/).

Here, a PVC pipe is connected, in order to permit the discharge of the chaff, either on the swath (Figure 3a) or onto the trailer (Figure 3b). The screw and the twin-stage turbine are driven by the dedicated hydraulic system.

The trailer used was a single-axed wagon, with a pivoted drawbar directly connected to the hitch of the combine (Figure 3a). The loading capacity of the trailer was 6 m^3. The upper part of the trailer was closed with a thick plastic film, in order to prevent accidental loss of chaff due to wind interference. The combine harvester was also equipped with auxiliary hydraulic connections, for controlling the movements of the trailer while discharging the chaff.

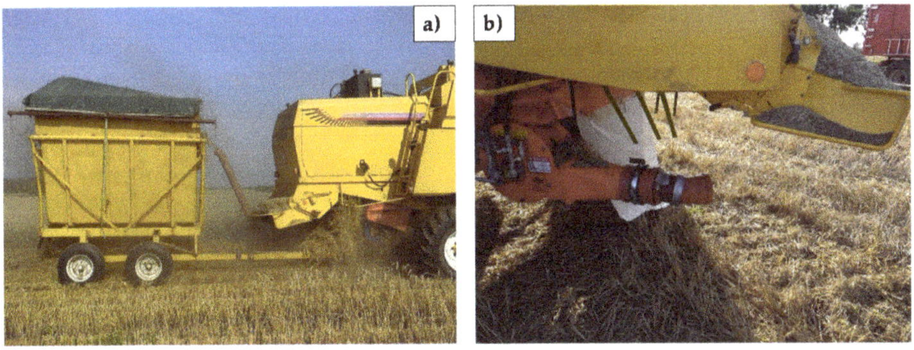

Figure 3. (**a**) Loose in a towed trailer (CoT) = chaff recovery system mounted in New Holland TX68, for the discharging of the product on a towed trailer (Trailer Agrohill Maskin AB, Halmstad, Sweden); (**b**) CoS = chaff recovery system mounted in New Holland TX68, for the discharging of the product on top of the swath.

2.3.2. Residual Biomass Harvesting

In treatment CoT, the chaff collected during the harvesting was systematically discharged into an auxiliary trailer parked outside the field, then weighted at the end of every plot, using a local scale. In both treatments, the straw were baled using a round baler John Deer 550 towed by a tractor John Deere 6830. The baler was completely empty at the beginning of each plot. At the end of each experimental unit, the machine was forced to close the bale, even if undersized. The last bale was included in the calculation of the residue production per plot, but not in the calculation of the mean weight of the bales, in order to avoid biased mean weights. In treatment CoS the straw swaths, that also included the chaff, were baled, according to the same methodology applied in CoT. In both treatments the fuel consumption registered by the on-board computer of the tractor was recorded for a fuel consumption calculation.

2.4. Harvesting and Baling Performance

Every plot guaranteed the formation of four swaths after harvesting, with minimal overlapping between the passes. In treatment CoT, the combine had to stop at least once, in order to empty the trailer and complete the harvesting. At the end of the plot, the trailer was emptied again for total chaff weight. The time required for discharge operations was recorded as accessory time. To measure the grain yield, the collected grain was discharged on a trailer and weighted for each plot.

The performance of the machines was evaluated through the study of the working times, performed according to the Comité International d'Organisation Scientifique du Travail en Agriculture (CIOSTA) methodology and the recommendations from the Italian Society of Agricultural Engineering (A.I.I.A.) 3A R1 [35]. The evaluation of the field speed allowed the determination of the theoretical field capacity (TFC, $ha \cdot h^{-1}$), the effective field capacity (EFC, $ha \cdot h^{-1}$), field efficiency (FE, %) and material capacity (MC, $t \cdot h^{-1}$). Gathered data were used to define the performance of the machines and the operative costs. Fuel consumption during baling was recorded by using the measuring system of the tractor. In the following paragraphs, the biomass unit (t) refers to fresh weight.

2.5. Post-Harvesting Test: Biomass Collected, Losses and Bulk Density

After baling, all bales produced within the plots were weighed singularly for total biomass baled assessment and average fresh weight measurement (here, the last bale was not included in the calculation). In treatment CoT, the quantity of the chaff collected was determined by weighing the chaff collected in each plot on an in situ scale.

Losses of biomass were assessed for stubble, straw and chaff. By knowing the cutting height of the combine header, stubbles were reconstructed in the laboratory by cutting the basal part of culms previously harvested for pre-harvest analysis. Straw and chaff losses were determined as the difference between the theoretical biomass available derived from the pre-harvest analysis and the effective biomass weighted at the end of the test. The moisture content of each biomass fraction was measured according to the standard methodology described above. The bulk density ($kg \cdot m^{-3}$) of the loose biomass stored in the trailer was assessed by taking 10 randomly selected samples of chaff and was measured according to ISO 17828:2015 [36]. In each plot, all bales were weighed singularly, and three of them were randomly selected and their sizes measured for volume assessment. Bulk density was successively calculated by dividing the mass in kilograms by the volume in cubic meters.

2.6. Cost Analysis

In the economic analysis, the following parameters have been taken into account: purchase and operating costs that were provided by the contractor via a interview, performance of the machines derived from the field tests as primary data, and standard values reported in CRPA methodology [35]. Hourly costs of machines were calculated on the basis of the market value of the agricultural machinery [37,38]. The prices of the machines have been discounted to 2019, applying the lending rate of 3% provided by Banca d' Italia Institute [39]. The parameters used during the cost analysis are reported in Tables 1 and 2.

Table 1. Parameters used for the economic analysis in CoT treatment. Harvesting stage with the collection of chaff on a towed trailer and straw baling stage.

		Unit	Harvesting			Baling	
Machine			Combine harvester			Tractor	
	Power	[kW]	208.8			115.6	
	Operating machine			Thierart	Trailer		Baler
Financial cost	Investment	[€]	230,980	10,000	7000	110,127	30,463
	Service life	[y]	10	10	10	10	8
	Service life	[h]	3000	3000	3000	14,000	2500
	Resale	[%]	19	18	18	32	23
	Resale	[€]	44,139	1768	1238	40,524	6878
	Depreciation	[€]	186,841	8232	5762	69,603	23,585
	Annual usage	[h·y^{-1}]	480	480	480	307	307
	Interest rate	[%]	3	3	3	3	3
	Workers	[n]	1	-	-	1	-
Fixed costs	Ownership costs	[€·y^{-1}]	18,684.09	823.16	576.21	11,009.49	2948.11
	Interests	[€·y^{-1}]	4126.79	176.53	123.57	1652.39	560.12
	machine shelter	[m^2]	62.30	0.00	10.20	9.12	6.93
	Value of the shelter	[€·m^{-2}]	100.00	0.00	100.00	100.00	100.00
	Value of the shelter	[€·y^{-1}]	124.59	0.00	30.60	27.36	20.79
	Insurance (0.25%)	[€·y^{-1}]	577.45	0.00	17.50	275.32	76.16
Variable costs	Repair factor	[%]	40.00	45.00	80.00	80.00	90.00
	Repairs and maintenance	[€·h^{-1}]	49.28	2.40	2.99	1.38	10.78
	Fuel cost	[€·l^{-1}]	0.57			0.57	
	Fuel consumption	[L·h^{-1}]	32.51			11.60	
	Fuel cost	[€·h^{-1}]	18.66			6.66	
	Lubricant cost	[€·l^{-1}]	3.03			3.03	
	Lubricant consumption	[L·h^{-1}]	0.14			0.09	
	Lubricant consumption	[€·h^{-1}]	0.44			0.27	
	Worker salary	[€·h^{-1}]	11.50			11.50	
	Cost of baling string	[€·h^{-1}]					32.32

Table 2. Parameters used for the economic analysis in baled together with the straw (CoS) treatment. The chaff collected with the twin-stage turbine is discharged on the swath and baled afterward.

		Unit	Harvesting		Baling	
Machine			Combine harvester		Tractor	
	Power	[kW]	208.8		115.6	
	Operating machine			Thierart		Baler
Financial cost	Investment	[€]	230,980	10,000	110,127	30,463
	Service life	[y]	10	10	10	8
	Service life	[h]	3000	3000	14,000	2500
	Resale	[%]	19	18	32	23
	Resale	[€]	44,139	1768	40,524	6878
	Depreciation	[€]	186,841	8232	69,603	23,585
	Annual usage	[h·y^{-1}]	480	480	307	307
	Interest rate	[%]	3	3	3	3
	Workers	[n]	1		1	
Fixed costs	Ownership costs	[€·y^{-1}]	18,684.09	823.16	11,009.49	2948.11
	Interests	[€·y^{-1}]	4126.79	176.53	1652.39	560.12
	Machine shelter	[m^2]	62.30		9.12	9.89
	Value of the shelter	[€·m^{-2}]	100.00		100.00	100.00
	Value of the shelter	[€·y^{-1}]	124.59		27.36	29.67
	Insurance (0.25%)	[€·y^{-1}]	577.45		275.32	76.16

Table 2. *Cont.*

		Unit	Harvesting		Baling	
Variable costs	Ownership costs	[%]	40.00	45.00	80.00	90.00
	Repairs and maintenance	[€·h^{-1}]	49.28	2.40	1.38	10.78
	Fuel cost	[€·L^{-1}]	0.57		0.57	
	Fuel consumption	[L·h^{-1}]	32.51		10.7	
	Fuel cost	[€·h^{-1}]	18.66		6.14	
	Lubricant cost	[€·L^{-1}]	3.03		3.03	
	Lubricant consumption	[L·h^{-1}]	0.14		0.09	
	Lubricant consumption	[€·h^{-1}]	0.44		0.27	
	Worker salary	[€·h^{-1}]	11.50		11.50	
	Cost of baling string	[€·h^{-1}]				32.32

In the calculation of the operating costs of the two harvesting systems, the time required for each operation, the quantity of the products obtained and the respective market value (Table 3) were considered. The economic allocation in each treatment was derived from the ratio between each product revenue on the total revenues obtained, as shown in the following formula:

$$Ea = \frac{Mp \times Y_i}{\sum_{i=1}^{3} R_i} \tag{1}$$

where:

Ea = Economic allocation of each product or co-product (i.e., grain seed, straw, or chaff) per harvesting phase (combine harvesting or baling)
Mp = Market price of each product or co-product (i.e., grain seed, straw, or chaff)
Y_i = Yield of each product or co-product (i.e., grain seed, straw, or chaff)
R_i = Revenue obtained by multiplying $Mp \times Y_i$

Table 3. Economic allocation used for the cost analysis of straw and chaff harvesting with the Thierart technology in Sweden, for each harvesting phase, and treatment.

Treatment	Product	Market Price [€·t^{-1}]	Yield [t·ha^{-1}]	Revenue [€·ha^{-1}]	Economic Allocation Harvesting [%]	Economic Allocation Baling [%]
CoT	Grain	198.5 [1]	9.83	1951.26	89.2	0.0
	Straw	50 [1]	3.88	194	8.9	100.0
	Chaff	50 [1]	0.84	42	1.9	0.0
	Total			2187.26	100.0	100.0
CoS	Grain	198.5	9.5	1881.78	88.9	0.0
	Straw	50	3.9	194	9.2	83.0
	Chaff	0	0.8	40	1.9	17.0
	Total			2115.78	100.0	100.0

Note: prices retrieved from Camera di Commercio di Modena (2019) [40].

2.7. Statistical Analysis

The statistical analysis was performed in order to discriminate the differences among the treatments. All data were subjected to the analysis of variance (ANOVA), using the R 3.6.1 to separate statistically different means ($p \leq 0.05$) [41].

3. Results and Discussions

3.1. Biomass Fractions

The results of pre-harvesting highlighted that the total aboveground biomass was 18.8 t·ha^{-1}. Spikes represented the 57% (10.78 t·ha^{-1}) of the total (47% seeds and 10% chaff, corresponding to 8.94 t·ha^{-1} and 1.84 [t·ha^{-1}], respectively) whereas the whole culms accounted for the 43% (8.02 t·ha^{-1}). The moisture content was equal to 14.3% (±9.1), 9.0% (±3.5) and 14.6% (±2.7), for straw, chaff and seeds, respectively. After the harvesting, the different fractions of biomass collected are shown in Figure 4.

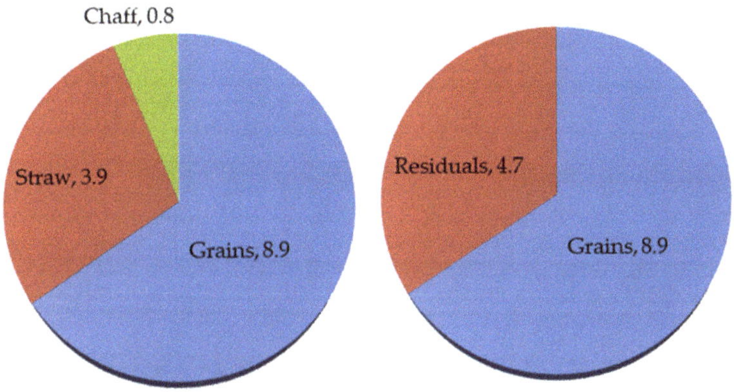

Figure 4. Effective tons of fresh biomass collected in treatment CoT (**left**) and CoS (**right**). CoT permitted one to collect the chaff separately from the straw, while in CoS, the chaff was baled with straw and considered as residual.

3.2. *Performance of the Combine*

The methodologies studied for chaff collection highlighted some differences in the performance of the machines involved. According to what was anticipated by Glasner et al. (2019) [42], the theoretical field capacity (TFC) of the combine did not variate among the treatments, as its speed was constant during the cutting and cleaning processes, although a reduction of 10–25% of cleaning was reported in the study. Despite that, significant differences were found in EFC, FE and MC (Table 4), where the accessory times, like the time required for unloading the wagon in CoT, were included. In fact, the wagon could collect only 6 m^3 of loose chaff and, considering the low bulk density of 41.75 kg·m^{-3}, the wagon shortly became full of chaff, forcing the combine harvester to stop and exit the field for unloading the wagon. A similar value of 42.88 kg·m^{-3} for chaff bulk density was reported by Bergonzoli et al. [26] and slightly higher values of 56 kg·m^{-3} and 62.08 kg·m^{-3} were found by McCartney et al. [13] and by Suardi et al. [24].

Table 4. Comparison of the performance of the combine harvester among the two treatments. (TFC = Theoretical Field Capacity, EFC = Effective Field Capacity, FE = Field Efficiency, MC = Material Capacity).

Treatment	TFC	EFC	FE	MC	
	[ha·h^{-1}]	[ha·h^{-1}]	[%]	[t_{grain}·h^{-1}]	[$t_{residue}$·h^{-1}]
CoT	2.17 ± 0.20	1.02 ± 0.18	47 ± 5	9.87 ± 0.82	0.84 ± 0.05
CoS	2.34 ± 0.43	1.71 ± 0.36	73 ± 8	16.22 ± 3.49	-
ANOVA	ns	*	**	*	

Note: (ns) not significant; (*) Significant at $p < 0.05$; (**) Significant at $p < 0.01$.

In CoT, the unproductive times of the combine harvester were 233% higher than in CoS where the chaff was continually blown over the swath. In fact, as depicted in Table 4, the FE and the MC of the combine harvester were significantly higher when the CoS system was applied. In Suardi et al. [19], TFC and EFC were respectively 3.72 ha·h^{-1} and 2.28 ha·h^{-1} on average and the combine fuel consumption resulted in 11.8 L·h^{-1}.

Different methods for loose chaff collection have been reported in the literature. For instance, Suardi et al. [24] tested a continuative discharging of chaff onto a trailer towed by a tractor side by side the combine; that the system permitted to collect 1.27 t·ha^{-1} of loose chaff. Differently, Bergonzoli et al. [26] tested a combine harvester equipped with Harcob system, which had an integrated tank of 9 m^3 in volume for storage of the chaff collected and that the system allowed to collect 0.6 t·ha^{-1}. Regardless of the quantity of the chaff collected, neither of them reported negative impacts on the combine performance: the trailer volume available for chaff storage in Suardi et al. [24] was better dimensioned, while the Harcob system allows the simultaneous discharging of grain and chaff, avoiding extra unloading time [26]. For that reason, the unproductive times needed were much lower. Similar tests performed by INRA (Institut National de la Recherche Agronomique) in the frame of the project «Systèmes de Cultures Innovants» and CUMA (Federation Nationale des Cooperatives d'Achat et d' Utilisation de Materiel Agricole) in 2011 and 2012 demodays with similar turbine systems, provided higher results in terms of quantity of chaff collected: respectively, 1.5 t·ha^{-1} and 1.15 t·ha^{-1} [43,44].

3.3. Performance of the Baler

Regarding the baling stage, the EFC that includes accessories' times (e.g., turning time and unloading time) was lower in CoS, since a higher quantity of biomass in the swath was to be processed (Table 5). That implied more stops for the discharge of the bales and it also forced the tractor to reduce the speed, in order to avoid overloading of the baler's chamber. In fact, the amount of biomass that the baler could process per unit of time was not statistically different. No significant differences were found regarding TFC. The fuel consumption of the tractor was also recorded and referred to the unit of biomass baled. On average, 1.27 (±0.17) l of diesel fuel was required for each ton of straw baled, regardless of the presence or the absence of the chaff in the bales.

Table 5. Comparison of the performance of the baler within the two treatments. MC is calculated taking into account the overall quantity of residual biomass produced: straw and chaff together (TFC = Theoretical Field Capacity, EFC = Effective Field Capacity, FE = Field Efficiency, MC = Material Capacity). No statistical differences were found between treatments.

Treatment	TFC [ha·h^{-1}]	EFC [ha·h^{-1}]	FE [%]	MC [t·h^{-1}] ⸸
CoT	2.86 ± 0.43	1.92 ±0.09	68 ± 7	7.47 ± 0.85
CoS	2.21 ± 0.25	1.57 ± 0.15	71 ± 2	7.37 ± 0.38
ANOVA	ns	*	ns	ns

Note: (⸸) Material Capacity refers to tons of fresh residues. (ns) Not significant; (*) Significant at $p < 0.05$.

Similar tests were performed by Suardi et al. in France in 2018 and 2019 [24], on baling the straw with chaff. The authors reported higher values for TFC, EFC and MC, respectively: 5.23 (±0.65) ha·h^{-1}, 3.46 (±0.28) ha·h^{-1} and 20.79 (±0.7) t·h^{-1} in 2018; whereas 4.64 (±0.31) ha·h^{-1}, 3.09 (±0.13) ha·h^{-1} and 20.20 (±2.0) t·h^{-1} in 2019. When chaff was not included in the bales, the performance of the baler was not statistically different. The fuel consumption ranged between 0.77 (±0.15) l·t^{-1} and 0.94 (±0.12) l·t^{-1} in the case of straw and chaff baling, while it ranged from 1.01 (±0.13) l·t^{-1} and 0.64 (±0.23) l·t^{-1}, when the chaff was dispersed on the ground. In similar experiment, the TFC and EFC of straw baling operation resulted on average 3.96 ha·h^{-1} and 2.01 ha·h^{-1}, with a mean FE of 51 % [19].

3.4. Losses of Biomass during the Baling Stage

The theoretical availability of straw, in the present study, was estimated in 8.02 t·ha^{-1}; in line with other studies such as Suardi et al. [24], where the theoretical straw availability was estimated at 7.39 (±0.73) t·ha^{-1} and 8.33 (±0.75) t·ha^{-1}, in 2018 and 2019 tests, respectively. Nevertheless, during the present study, the amount of residues baled was on average 3.88 t·ha^{-1} and 4.68 t·ha^{-1} with CoT and CoS treatments, respectively (Table 6). Therefore, the remarkable differences in the residue harvesting performance can be imputed exclusively to the suitability of the machine chosen by the contractor, to carry on the baling stage. The round baler John Deere mod. 550 used during the test was equipped with a pick-up 1.41 m wide, whereas the straw swath produced by the combine harvester measured 1.74 m in width, on average. Hence, 0.33 m of straw swath could not be collected by the baler's pick-up system in each pass, due to reduced width of the its pickup system (Figure 5). At the end of the baling phase, a large quantity of product was still not harvested in the field (Figure 3).

Table 6. Differences in fresh biomass outputs from wheat crop, due to the use of a twin-stage Turbine for chaff collection.

Treatment	Machine	Yield Grain [t·ha^{-1}]	 Residue [t·ha^{-1}]	Bale Weight [kg]	Bale Density [kg·m^{-3}]	Chaff Bulk Density [kg·m^{-3}]
CoT	Combine	9.83 ± 1.26	0.84 ± 0.12	-	-	41.75 ± 3.30
	Baling	-	3.88 ± 0.06	184.6 ± 4.41	76.78 ± 1.83	-
CoS	Combine	9.48 ± 0.45	-	-	-	-
	Baling	-	4.68 ± 0.23	198.4 ± 3.14	82.22 ± 1.33	-
ANOVA		ns	ns ⱡ	*	*	

Note: (ⱡ) In treatment CoT the mean residue value takes into account also the chaff, (-) not performed; (ns) non-significant; (*) Significant at $p < 0.05$.

Figure 5. The narrow pick-p of the round baler (**left**) caused high loss of straw (**right**) along the swath (areas of the swath not reached by the baler's pickup system are highlighted in red).

The estimated average loss of residue after baling was 4.75 t·ha^{-1} (4.68 t·ha^{-1} for CoS, and 4.72 t·ha^{-1} for CoT), corresponding to a loss of biomass of 50% on average, without statistical difference between the two treatments.

Bergonzoli et al. [26] reported a similar value when a combine harvester mounting Harcob system (developed for Maize cob harvesting) was modified and used for collecting the chaff in wheat crops, even if the results were ascribed to the cleaning system of the combine harvester.

Such a level of product losses recorded during the tests exceed the sustainable removal rate of 33% proposed by Unger and Glasner (2019) [23]. For this reason, it could be considered positive from the point of view of soil fertility, even if the economic sustainability is closely linked to the amount of recoverable biomass. Therefore, low collection efficiencies may render the operation of recovering residual biomass economically unviable.

However, the scenarios herein proposed provided differences in both the quantity and quality of residuals biomass collectable from wheat cropping, without affecting the grain yield. Such an aspect is very important; in fact harvesting, along with storage, is the most responsible factor for loss of grains throughout the wheat supply chain [45]. The presence of the chaff included in the bales increased both weight and density of the bales by 7.45% and 7.09% respectively, in comparison with bales free of chaff (Table 6). Increases of 18.0% in bale bulk density, due to the inclusion of chaff, was reported by Suardi et al. [24], when a similar turbine technology for chaff recovery was used. On the other hand, Suardi et al. reported a non-significant increase in the case of chaff admixing performed with a combi system (manifactured by Rekordverken Sweden AB, Kvänum, Sweden) [19].

The different methods studied, allowed to harvest 4.68 t·ha^{-1} and 4.72 t·ha^{-1} of wheat residues by baling chaff and straw together, or by harvesting chaff in the trailer and straw baling, respectively (Table 6), with no statistical differences.

3.5. Cost Analysis

In the analysis of the unitary costs, the running cost of each machinery involved in the supply chain is related to the market price [€·t^{-1}] of each product and by-product obtained. The performance of the machines contributed to the final calculation of the costs. For instance, the reduction in EFC, FE and MC of the combine harvester (Table 4) found that, when the combine towed the wagon (CoT), it increased the hourly harvesting cost by 3.41%, the cost per hectare by 73.35%, and the cost per ton of biomass processed by 67.73% (Tables 7 and 8), in comparison with CoS. Here, the combine harvester did not waste time to continually stop and unload the wagon.

Table 7. Costs for unit of time, surface and ton of biomass processed in CoS harvesting system, considering the productivity and the market price of each product.

		Unit	Grain	Straw	Chaff	Total Harvesting Costs
	Market price	[€·t^{-1}]	198.5	50	50	
	Yield	[t·ha^{-1}]	9.83	3.88	0.84	
Harvesting	Cost allocation	[%]	89%	9%	2%	100%
	Combine harvester + Twin stage turbine + Wagon	[€·h^{-1}]	123.01	12.23	2.65	137.89
		[€·ha^{-1}]	120.6	11.99	2.6	135.18
		[€·t^{-1}]	12.27	3.09	3.09	18.45
Baling	Cost allocation	[%]	0%	100%	0%	100%
	Tractor + Baler	[€·h^{-1}]		116.85		116.85
		[€·ha^{-1}]		60.86		60.86
		[€·t^{-1}]		15.69		15.69
Total cost of the harvesting system		[€·h^{-1}]	123.01	129.08	2.65	254.74
		[€·ha^{-1}]	120.6	72.85	2.6	196.04
		[€·t^{-1}]	12.27	18.78	3.09	

Table 8. Costs for unit of time, surface and ton of biomass processed in CoT harvesting system considering the productivity and the market price of each product.

		Unit	Grain	Straw	Chaff	Total Harvesting Costs
	Market price	[€·t^{-1}]	198.5	50	50	
	Yield	[t·ha^{-1}]	9.48	3.88	0.8	
Harvesting	Cost allocation	[%]	89%	9%	2%	100%
	Combine harvester + Twin stage turbine	[€·h^{-1}]	118.59	12.23	2.52	133.34
		[€·ha^{-1}]	69.35	7.15	1.47	77.98
		[€·t^{-1}]	7.32	1.84	1.84	11
Baling	Cost allocation	[%]		83%	17%	100%
	Tractor + Baler	[€·h^{-1}]		96.47	19.89	116.36
		[€·ha^{-1}]		61.45	12.67	74.12
		[€·t^{-1}]		15.84	15.84	31.67
Total cost of the harvesting system		[€·h^{-1}]	118.59	108.7	22.41	249.7
		[€·ha^{-1}]	69.35	68.6	14.14	152.09
		[€·t^{-1}]	7.32	17.68	17.68	

On the other hand, when the chaff was blown on the swath, the baler had much more biomass (straw and chaff) to process. In fact, the baler's EFC (Table 5) dropped by 18.23% and the costs per hectare and per ton of biomass processed increased by 21.79% and 101.85%, respectively. The hourly cost for baling did not change (Tables 7 and 8).

The choice to apply CoS over CoT harvesting method affected both the performance and running cost of the machines. According to Table 9, the harvesting cost per hectare increased by 28.90% (from 152.03 €·ha^{-1} to 196.05 €·ha^{-1}), when the chaff was collected as loose material (CoT).

Table 9. Economic calculation cost, revenue and the net gain obtained from the collection of grains, straw and chaff when harvested with the two different methods: CoS and CoT.

Treatment	Product	Yield [t·ha^{-1}]	Market Price [€·t^{-1}]	Revenue [€·ha^{-1}]	Harvesting Costs [€·ha^{-1}]	Net Gain [€·ha^{-1}]
CoT	Seed	9.83	198.50	1951.26	120.60	1830.66
	Straw	3.88	50.00	194.00	72.85	121.15
	Chaff	0.84	50.00	42.00	2.60	39.40
	Total	14.55	-	2187.26	196.05	1991.21
CoS	Seed	9.48	198.50	1881.78	69.35	1812.43
	Straw	3.88	50.00	194.00	68.60	125.40
	Chaff	0.80	50.00	40.00	14.14	25.86
	Total	14.16		2115.78	152.09	1963.69

The same results were obtained by Unger and Glasner in 2019, where the separate chaff collection and supply led to higher costs [23]. However, the overall capacity of CoT system permitted one to collect more biomass per hectare (0.38 t and 0.04 t of grain and chaff respectively), counterbalancing the higher costs. In fact, if considering just the net gain per hectare, CoS permitted to gain only 27.52 €·ha^{-1}.

Furthermore, in the present study, a market price for chaff of 50 €·t^{-1} was considered. However, Unger and Glasner [23] highlighted that the potential revenue of chaff could vary depending the final use and market price that can range from 81 €·t^{-1} to 200 €·t^{-1}, making chaff separate collection economically viable, and giving the farmer, from year to year, different sales opportunities of the product to more profitable markets.

4. Conclusions

The cultivation of the cereals is an important source of staple food around the world, and it also produces a relevant quantity of ligno-cellulosic biomass, that can be further exploited in order to improve the economic and environmental sustainability of the whole supply chain. In fact, agricultural residues are gaining more and more interest, due to their considerable availability and their potential content of energy, or as raw material for industrial processes. Cereal straw and chaff collected either separately or baled altogether can be a source of food for animals, particularly in case of shortage, or natural bedding for livestock. In poultry farming, farmers reported positive experiences on the use of loose chaff for littering, since it provides wellness to animals and a good adsorbent capacity. However, possible utilization of chaff is to produce bioenergy. Normally, about two tons of chaff per hectare are available, but still not collected, due to three major problems: unawareness of proper mechanical devices available on the market for its collection, uncertainty on the harvesting system to adopt and the development of a specific supply chain for its exploitation. So far, the literature reports few cases of chaff collection with the specific purpose of weed seeds removal, but it still lacks specific experiments on these machines intentionally used for biomass collection. Therefore, the present study aimed to fill that gap and provide deeper understanding in the possibility to enhance the current wheat harvesting method, in order to improve the quantity of biomass collected by including the chaff. This research analyzed the technical and economic feasibility of two different logistic methods for chaff collection: chaff collected as loose product onto a towed trailer (CoT) and baled altogether with the straw (CoS).

Our results suggest that upgrading a conventional combine harvester with a twin-stage turbine for chaff collection increases the total biomass collected by 0.84 $t{\cdot}ha^{-1}$ without affecting the grain yield. Furthermore, the separation of chaff from the straw is performed simultaneously to the cleaning process of the grain and no additional passes of the machine on the field are needed, and further soil compaction is prevented.

Even if our results reveal that the collection of the loose chaff into a towed wagon is more costly than including it into the bales, the market price of the pure chaff should be higher, to offset the extra costs required by the contractor for the collection and handling. Furthermore, it should be noted that the trailer system could be used also for other crop by-products; for instance, collecting finely chopped roughage after a forage harvester, without the strong modification of the combine, reducing the unitary cost of the investment and increasing the quantity of biomass potentially collectable. In fact, the unproductive time in CoT was 233% higher than in CoS with an increase of 43.94 $€{\cdot}ha^{-1}$ for the harvesting cost. In addition to the higher costs, loss of revenue may take place in case of inappropriate choice of the machine for accomplishing a specific task. Particularly, the round baler chosen by the contractor could not collect all the straw windrowed by the combine harvester. Although the subject is still under discussion, some authors consider that a residue extraction of no more than 33% is sustainable for the soil fertility. On the other hand, however, an amount of uncollected residue, such as that found during the study (50% of harvesting losses), could negatively affect the economic feasibility of the residue collection phase, questioning the investment in specific equipment. In fact, according to 6results from CoT treatment, where the chaff was not included in the straw, only 3.88 $t{\cdot}ha^{-1}$ out of 8.02 $t{\cdot}ha^{-1}$ of straw available on the field were baled. Considering the straw market price of 50 $€{\cdot}t^{-1}$, this can be translated into a loss of income of more than 200 $€{\cdot}ha^{-1}$.

Future studies should be focused on the assessment of the sustainability of the chaff collection, in terms of the effect to the soil fertility, carbon dioxide emissions and soil compaction.

Author Contributions: Conceptualization and methodology, A.S.; Investigation and data curation A.S., W.S., F.L.; writing—original draft preparation W.S., S.B.; writing—review and editing, A.S., L.P. and N.J.; supervision, L.P. and A.S.; funding acquisition, L.P. All authors have read and agreed to the published version of the manuscript.

Funding: This research was funded by European Union's Horizon 2020 AGROinLOG project grant number 727961 (http://agroinlog-h2020.eu/en/home/).

Acknowledgments: Authors thank the contractor Leif Jönsson (Tjärby Gästgivaregård Laholm, Sweden) and his team for their valid support and assistance provided during the activities, as well as Sandu Lazar for their valuable contribution in the field activities. Moreover, the authors thank Consuelo Attolico for looking after the relationship with the French company ETS Thierart®.

Conflicts of Interest: The authors declare no conflict of interest.

References

1. Eise, J.; Foster, K. *How to Feed World*; Island Press Center for Resource Economics: Washington, DC, USA, 2018; pp. 1–250.
2. European Parliament. DIRECTIVE (EU) 2015/1513 of the european parliament and of the council of 9 September 2015 amending Directive 98/70/EC relating to the quality of petrol and diesel fuels and amending Directive 2009/28/EC on the promotion of the use of energy from renewabl. *Off. J. Eur. Union* **2015**, *L239/1*, 20–30.
3. EU. A 2030 Framework for Climate and Energy Policies. European Parliament Resolution of 5 February 2014 on a 2030 Framework for Climate and Energy Policies (2013/2135(INI)). 2014. Available online: https://eur-lex.europa.eu/legal-content/EN/TXT/PDF/?uri=CELEX:52014IP0094&from=EN (accessed on 18 January 2020).
4. Stelte, W.; Sanadi, A.R.; Shang, L.; Holm, J.K.; Ahrenfeldt, J.; Henriksen, U.B. Recent developments in biomass pelletization—A review. *BioResources* **2012**, *7*, 4451–4490.
5. Paiano, A.; Lagioia, G. Energy potential from residual biomass towards meeting the EU renewable energy and climate targets. The Italian case. *Energy Policy* **2016**, *91*, 161–173. [CrossRef]
6. Mann, M.E.; Cohen, R.D.H.; Kernan, J.A.; Nicholson, H.H.; Christensen, D.A.; Smart, M.E. The feeding value of ammoniated flax straw, wheat straw and wheat chaff for beef cattle. *Anim. Feed Sci. Technol.* **1988**, *21*, 57–66. [CrossRef]
7. Ionel, I.; Cioablă, A.E. Biogas production based on agricultural residues. From history to results and perspectives. *WSEAS Trans. Environ. Dev.* **2010**, *6*, 591–603.
8. Fang, G.; Shen, K. Wheat straw pulping for paper and paperboard production. In *Global Wheat Production*; InTech Open: London, UK, 2018; pp. 223–240.
9. Hýsek, Š.; Čermák, J.; Lexa, M. Influence of lignocellulosic waste pre-treatment on the characteristics of bond rupture. *Sustainability* **2019**, *11*, 4784. [CrossRef]
10. Swain, M.R.; Singh, A.; Sharma, A.K.; Tuli, D.K. Bioethanol production from rice—And wheat straw: An overview. In *Bioethanol Production from Food Crops*; Ray, R.C., Ramachandran, S., Eds.; Elsevier: Amsterdam, The Netherlands, 2019; pp. 213–231, ISBN 978-0-12-813766-6.
11. Yang, Y.; Ma, S.; Zhao, Y.; Jing, M.; Xu, Y.; Chen, J. A field experiment on enhancement of crop yield by rice straw and corn stalk-derived biochar in Northern China. *Sustainability* **2015**, *7*, 13713–13725. [CrossRef]
12. Seglah, P.A.; Wang, Y.; Wang, H.; Bi, Y. Estimation and efficient utilization of straw resources in Ghana. *Sustainability* **2019**, *11*, 4172. [CrossRef]
13. McCartney, D.H.; Block, H.C.; Dubeski, P.L.; Ohama, A.J. Review: The composition and availability of straw and chaff from small grain cereals for beef cattle in western Canada. *Can. J. Anim. Sci.* **2006**, *86*, 443–455. [CrossRef]
14. EU. Eurostat Database, Cereals for the Production of Grain (Including Seed) by Area, Production and Humidity. 2019. Available online: https://ec.europa.eu/eurostat/databrowser/view/tag00027/default/table?lang=en (accessed on 18 January 2020).
15. Rönnbäck, M.; Lundin, G. Simultaneous harvesting of straw and chaff for energy purposes-influence on bale density, yield, field drying process and combustion characteristics. In Proceedings of the XVIIth World Congress of the International Commission of Agricultural and Biosystems Engineering (CIGR) Hosted by the Canadian Society for Bioengineering (CSBE/SCGAB), Québec City, QU, Canada, 13–17 June 2010.
16. Weiß, B.D.; Glasner, C. Evaluation of the process steps of pretreatment, pellet production and combustion for an energetic utilization of wheat chaff. *Front. Environ. Sci.* **2018**, *6*, 1–10. [CrossRef]
17. Zhang, H.; Lopez, P.C.; Holland, C.; Lunde, A.; Ambye-Jensen, M.; Felby, C.; Thomsen, S.T. The multi-feedstock biorefinery—Assessing the compatibility of alternative feedstocks in a 2G wheat straw biorefinery process. *GCB Bioenergy* **2018**, *10*, 946–959. [CrossRef]

18. Shirtliffe, S.J.; Entz, M.H. Chaff collection reduces seed dispersal of wild oat (avena fatua) by a combine harvester. *Weed Sci.* **2005**, *53*, 465–470. [CrossRef]
19. Uardi, A.; Saia, S.; Stefanoni, W.; Gunnarsson, C.; Sundberg, M.; Pari, L. Admixing chaff with straw increased the residues collected without compromising machinery efficiencies. *Energie* **2020**, *13*, 1766.
20. Walsh, M.; Newman, P.; Powles, S. Targeting weed seeds in-crop: A new weed control paradigm for global agriculture. *Weed Technol.* **2013**, *27*, 431–436. [CrossRef]
21. Jacobs, A.; Kingwell, R. The harrington seed destructor: Its role and value in farming systems facing the challenge of herbicide-resistant weeds. *Agric. Syst.* **2016**, *142*, 33–40. [CrossRef]
22. Walsh, M.J.; Harrington, R.B.; Powles, S.B. Harrington seed destructor: A new nonchemical weed control tool for global grain crops. *Crop Sci.* **2012**, *52*, 1343–1347. [CrossRef]
23. Unger, J.S.; Glasner, C. Cost analysis of chaff harvesting concepts in Germany. *Agronomy* **2019**, *9*, 579. [CrossRef]
24. Suardi, A.; Stefanoni, W.; Alfano, V.; Bergonzoli, S.; Pari, L. Equipping a combine harvester with turbine technology increases the recovery of residual biomass from cereal crops via the collection of chaff. *Energies* **2020**, *13*, 1572. [CrossRef]
25. Pari, L.; Toscano, G.; Suardi, A.; Bergonzoli, S.; Lopez, E.; Scarfone, A.; Alfano, V. Maize cob and cereal chaff: Feedstocks for energy production. *Eur. Biomass Conf. Exhib. Proc.* **2018**, *2018*, 279–282.
26. Bergonzoli, S.; Suardi, A.; Rezaie, N.; Alfano, V.; Pari, L. An innovative system for maize cob and wheat chaff harvesting: Simultaneous grain and residues collection. *Energies* **2020**, *13*, 1265. [CrossRef]
27. Gerling, M.; Dickey, P.C. Absorbant Animal Bedding. U.S. Patent No 5,878,696, 9 March 1999. Available online: https:/patentimages.storage.googleapis.com/aa/1d/1d/5d9a9f2a556454/US5878696.pdf (accessed on 25 February 2020).
28. Anisuzzaman, M.; Chowdhury, S.D. Use of four types of litter for rearing broilers. *Br. Poult. Sci.* **1996**, *37*, 541–545. [CrossRef] [PubMed]
29. Swain, B.K.; Sundaram, R.N.S. Effect of different types of litter material for rearing broilers. *Br. Poult. Sci.* **2000**, *41*, 261–262. [CrossRef] [PubMed]
30. Shishkin, V.; Shulzhenko, E. The study of soybean chaff pressing process depending on its fractional composition and humidity. *Bull. Sci. Pract.* **2019**, *5*, 160–164. [CrossRef] [PubMed]
31. Trinh Van, Q.; Nagy, S. The influence of temperature on the briquetting of ground post agglomerated spelt chaff. *Mech. Hulladékkezeles* **2019**, *13*, 13–18.
32. EU. Agroinlog—Integrated Biomass Logistic Centres fo the Agro-Industry. Available online: https://ec.europa.eu/eip/agriculture/en/news/agroinlog-integrated-biomass-logistics-centres (accessed on 26 May 2020).
33. Thiérart Etude. Conception et Réalisation de Machines Spéciales. Available online: https://www.thierart.fr/ (accessed on 17 January 2020).
34. NSAI. ISO. ISO 18134-2:2017 Solid Biofuels—Determination of Moisture Content—Oven Dry Method—Part 2: Total Moisture—Simplified Method 2017. Available online: https://www.iso.org/standard/71536.html (accessed on 6 March 2020).
35. Assirelli, A.; Croce, S.; Acampora, A.; Civitarese, V.; Suardi, A.; Santangelo, E.; Pari, L. An innovative system for conditioning biomass [sorghum bicolo (l.) moench.]. *Am. Soc. Agric. Biol. Eng.* **2013**, *56*, 829–837.
36. ISO 17828:2015 Solid Biofuels—Determination of Bulk Density. 2015. Available online: https://www.iso.org/obp/ui/#iso:std:iso:17828:ed-1:v1:en (accessed on 23 December 2015).
37. American society of agricultural engineers agricultural machinery management. *Agricultural Machinery Management*; American Society of Agricultural Engineers: St. Joseph, MI, USA, 2000; pp. 344–349.
38. Assirelli, A.; Pignedoli, S. Costo di esercizio delle macchine agricole. *Boll. CRPA Not.* **2005**, *5*, 1–10.
39. Banca d'Italia. Banca d'Italia Lending Rate. Available online: https://www.bancaditalia.it/ (accessed on 11 November 2019).
40. Camera di Commercio di Modena (Chamber of Commerce of Modena). Listino dei prezzi all'ingrosso rilevati sul mercato di Modena nella settimana dal 16 luglio al 22 luglio 2019 (List of Wholesale Prices Recorded on the Modena Market in the Week from 16 July to 22 July 2019). 2019. Available online: https://www.mo.camcom.it/tutela-del-mercato/borsamerci/listini/listino-dei-prezzi-allingrosso-rilevati-sul-mercato-di-modena-nella-settimana-dal-16-al-22-luglio-2019 (accessed on 6 March 2020).
41. R Development Core Team. *R: A Language and Environment for Statistical Computing*; R Foundation for Statistical Computing: Vienna, Austria, 2008; ISBN 3-900051-07-0.

42. Glasner, C.; Vieregge, C.; Robert, J.; Fenselau, J.; Bitarafan, Z.; Andreasen, C. Evaluation of new harvesting methods to reduce weeds on arable fields and collect a new feedstock. *Energies* **2019**, *12*, 1688. [CrossRef]
43. INRA. Recuperateur de Menues Pailles «turbo Paille de Thievin». Mignaloux-Beauvoir. 2011. Available online: https://geco.ecophytopic.fr/documents/20182/21720/upload_00011189_pdf (accessed on 6 March 2020).
44. CUMA. Journée Technique: Récuperation des Menues Pailles. 2015. Available online: http://draaf.normandie.agriculture.gouv.fr/IMG/pdf/6_Powerpoint_journee_menues_pailles_FRCUMA_cle0bd622.pdf (accessed on 6 March 2020).
45. Mesterházy, A.; Oláh, J.; Popp, J. Losses in the grain supply chain: Causes and solutions. *Sustainability* **2020**, *12*, 2342. [CrossRef]

sustainability

MDPI

Article

Postharvest Losses of Pomegranate Fruit at the Packhouse and Implications for Sustainability Indicators

Ikechukwu Kingsley Opara [1], Olaniyi Amos Fawole [1,2] and Umezuruike Linus Opara [1,*]

1 SARChI Postharvest Technology Research Laboratory, Africa Institute for Postharvest Technology, Faculty of AgriSciences, Stellenbosch University, Stellenbosch 7600, South Africa; ikekings101@gmail.com (I.K.O.); olaniyif@uj.ac.za (O.A.F.)

2 Postharvest Research Laboratory, Department of Botany and Plant Biotechnology, University of Johannesburg, Johannesburg 2006, South Africa

* Correspondence: opara@sun.ac.za; Tel.: +27-21-808-4064

Abstract: Pomegranate fruit, like other types of fresh horticultural produce, are susceptible to high incidence preharvest and postharvest losses and waste. Several studies have been done to improve the production and handling of pomegranate fruit to meet market standards, but little has been done in loss quantification, especially in the early stage of the value chain such as the packhouse. Therefore, the aim of this study was to quantify the magnitude of pomegranate fruit losses at the packhouse, identify the causes, and estimate the impacts of losses. The study was conducted on a case study packhouse in the Western Cape Province of South Africa from February to March 2020. The direct measurement method, which involved physical identification of the causes of loss on individual fruit, was used for data collection. Loss quantification involved the calculation of lost fruit proportional to the amount put in the packhouse processing line. The results showed that losses ranged between 6.74% to 7.69%, which translated to an average of 328.79 tonnes of pomegranate fruit removed during packhouse operation per production season at the investigated packhouse. This magnitude of lost fruit was equivalent to over ZAR 29.5 million (USD 1,754,984) in revenue, in addition to the opportunity costs of resources used to produce lost fruit.

Keywords: pomegranate; losses; nutrition; environmental; resources; packhouse; postharvest; impacts

Citation: Opara, I.K.; Fawole, O.A.; Opara, U.L. Postharvest Losses of Pomegranate Fruit at the Packhouse and Implications for Sustainability Indicators. *Sustainability* **2021**, *13*, 5187. https://doi.org/10.3390/su13095187

Academic Editors: Alessandro Suardi and Nadia Palmieri

Received: 13 February 2021
Accepted: 26 April 2021
Published: 6 May 2021

1. Introduction

Pomegranate (*Punica granatum L.*) is an ancient fruit believed to be first cultivated around 3000 and 4000 BC, and was mentioned both in the Bible and the Quran [1]. Its origin is traced to the Middle East, in present-day Iran, and it adapts to a variety of soil conditions in the Mediterranean, subtropical, and tropical climates [2,3]. Currently, it is grown in many countries for fresh consumption and industrial uses [4,5]. As a result, more than 500 cultivars are grown globally, with some cultivars named differently in different parts of the world [2,4–6]. The awareness of its numerous uses and benefits has made it popular among other fruit [5,6]. Pomegranate can be eaten as fresh produce or juiced and stored in the appropriate temperature and relative humidity. It is sweet, sour, or acidic depending on the cultivar and rich in vitamins, minerals, and other organic compounds [4,5]. The consumption of pomegranate has been linked with a great health outcome in different studies [3,7–10]. The phenolic compounds present in pomegranate have been found to be great anti-inflammatory, anti-oxidative, and anti-carcinogenic chemical compounds, which helps to reduce tumour growth and chronic inflammation [11]. The hypoglycaemic activity of pomegranate juice has been found to prevent diabetes mellitus [12]. Pomegranate fruit consumption has been reported to reduce cardiovascular diseases [13]. Chemical compounds in pomegranate fruit are also used in the treatment of diseases such as ulcers, acidosis, haemorrhage, aphthae, diarrhoea, dysentery, respiratory pathologies, and microbial infections [3,9]. The manufacturing industries use pomegranate aril and peel as a raw material in the production of jams, ink, dye, and oil [3].

Trends in Production and Trade of Pomegranate Fruit in South Africa and Globally

There has been a rapid increase in the production of pomegranate globally, but the trade has grown more locally in the major producing countries [14,15]. Countries such as India, China, Iran, and Turkey are the leading producers, while India and Iran are the highest exporters of the fruit [2]. However, because the fruit are often grown and picked from small farms in different locations in the major producing countries, there are no articulated data available about global production area [1,2,15]. However, global production was estimated to have increased from about 3 million tonnes in 2014 to 3.8 million tonnes in 2017 [1].

The production of pomegranate fruit in different parts of the world is primarily divided into two (Northern Hemisphere and Southern Hemisphere) due to different seasons of production in the regions [15,16]. The demand for the fruit, especially in the Northern Hemisphere, is derived in nature in the sense that it is mainly driven by industrial usage since there is no close substitute for the antioxidants found in pomegranate [15]. The supply is stratified according to the variation in the production seasons, which allows the Southern Hemisphere to fill the niche market gap in the Northern Hemisphere. The Northern Hemisphere, however, accounts for above 90% of the total production, hence, it has a higher share in the global trade [15]. According to Kahramanoglu [1], global pomegranate production area is increasing, but some of the producing countries are facing quality issues, which leads to considerable postharvest losses and waste. Europe is the biggest market for pomegranate followed by Asia and the Middle East, as almost all the producing countries share the European markets [1]. Peru and Chile are the biggest exporters of pomegranate fruit from the Southern Hemisphere with 74% and 14% respectively, while South Africa and Argentina have a combined 12% contribution to export from the region [17]. Iran, China, India, Turkey, Spain, and Israel are the highest producers in the Northern Hemisphere, but most of the production in this region is consumed locally [1,14].

The pomegranate fruit is one of the deciduous fruits grown in South Africa, occupying about 1024 hectares of land in 2019 from 771 hectares in 2011 [16]. It is mostly grown in the Western Cape, which accounts for about 81% of total production [16]. In South Africa, the majority of the production is exported, which earns export revenue for the country and income for the farmers and value chain actors. In 2019, about 76% of the total production was exported [16], and in 2018, the local market generated about ZAR 67,000 per tonne [17]. Production and export have grown from about 837,250 cartons (3.8 kg equivalent) in 2014 to 1,676,160 cartons (3.8 kg equivalent) in 2019, and are projected to increase to 2,055,271 cartons (3.8 kg equivalent) by 2024 [16]. Between 2014 and 2019, about 7,557,906 cartons (3.8 kg equivalent) were exported [16]. The major market for South African pomegranate is Europe. About 61% of the total export is in the European markets and 22% in the Middle East, with Asian and African countries importing a small amount [17]. The main competition for export market share comes from the Southern Hemisphere countries, whose pomegranate fruit is ready in the market almost in the same period as that of South Africa.

While the pomegranate industry is growing rapidly in South Africa and globally, fruit are susceptible to losses and waste (wastage) which reduce profitability due to a wide range of preharvest and postharvest factors, including pest and disease attack [18,19], bruise damage [20], moisture loss [21,22], and mechanical damage [23]. Industry estimates in South Africa also suggest that the incidence sunburn (a preharvest skin defect) alone can be high, causing grower losses that may exceed 30% of harvested fruit [24]. Despite the identified causes of wastage of the fruit in South Africa, there is a lack of quantitative and science-based data on the magnitude of losses to guide the implementation of loss reduction strategies. During typical packhouse operation, fruit are cleaned, sorted, graded, labelled, and packed, and those that do not meet quality specifications due to the preharvest and postharvest factors outlined are considered as loss and thus discarded or sold at a nominal price for juice or animal feed. As the last point of fruit handling and quality control prior to storage, marketing, and distribution, the packhouse is a critical step in assessing the magnitude and causes of fruit postharvest loss which are critical pieces of information for informing loss reduction strategies. Discussions on the magnitude of postharvest losses and the causes are often based on estimates, without site-level measurements which are known to be difficult and costly. However, researchers globally agree on the need for more studies to directly quantify the amount of postharvest food losses and to identify the site-specific contributing factors along the value chain [25–28]. Therefore, the objective of the current research was to assess the magnitude of pomegranate fruit postharvest losses at the packhouse level based on a case study in South Africa, identify the causes, and estimate the socio-economic and environmental impacts of the losses.

2. Materials and Methods

2.1. Study Settings

The study was conducted from February to March 2020 in the Western Cape Province (Latitude 33.2278° S, Longitude 21.8569° E) of South Africa. The province was chosen because over 80% of total production of pomegranate fruit in the country is done in the region [16] and the case study packhouse is arguably the biggest packhouse in the area. The study was conducted on the three most commercially grown pomegranate cultivars in the country, namely 'Herskawitz', 'Acco', and 'Wonderful'. 'Herskawitz' has a sour taste with hard seeds, 'Acco' has a sweet taste with softer seeds, while 'Wonderful' has a vinous taste with soft seeds [29]. The study was carried out by assessing the physical quality of fruit sorted as 'waste' from the packhouse production line. The assessment started at about 9:00 a.m. and ended by 3:00 p.m. daily. 'Herskawitz', which is the early cultivar, was assessed by mid-February while 'Acco' was assessed by early March. 'Wonderful' was assessed by mid and late March. The handling and packaging practices at the packhouse were observed. The unit of measurement of lost fruit was the bin [length (1270 mm) × width (1070 mm) × height (720 mm)]. A total of 251 bins, containing 1300–1500 fruit each which were processed by the packhouse during the study period, were used to assess the magnitude of fruit losses ('Herskawitz' 84, 'Acco' 89, and 'Wonderful' 78) by putting fruit through the packhouse line for sorting and grading. The bins handled by the packhouse during the study period were all examined in order to provide a sufficient and representative dataset based on commercial practice. The magnitude of loss was estimated for each pomegranate cultivar based on the number of waste bins containing lost/rejected fruit. Fruit sorted into waste bins were sampled into ventilated cartons (length (35 mm) × height (25 mm) × width (22 mm)) and later examined individually to determine the causes of loss. A total of 18 bins containing lost (discarded) fruit, six bins for each cultivar, were used to determine the causes of fruit loss by examining each fruit individually. Given that the same person carried out all the individual fruit assessment to reduce human error, this was the maximum number of bins and fruit that could possibly be examined during the research period. Loss calculations included the number of bins of discarded fruit for defect reasons proportional to the number of bins put in the processing line. The assessment was made based on the external quality of fruit. Quantification was done by collecting sample fruit to identify reasons for loss (defects) and how they contribute to total fruit loss.

2.2. Method of Data Collection

The research method for this study was the sampling method, which has been identified as a practical method for conducting a study where there is a large variable of data to consider and also in conditions where data collection is constrained [30]. Because the assessment of this present study was carried out simultaneously during full packhouse commercial operations, which constrained space and time for data collection, it was necessary to use the sampling method. Researchers have used sampling methods to conduct postharvest studies [31–33]. This present study involved the physical identification of the causes of fruit loss by examining individual fruit sorted into the waste bin at the packhouse. Qualitative data were also collected by physical observation during packhouse operation and interaction with the packhouse workers.

The economic impact of fruit losses was estimated using the supermarket retail price (ZAR 89.99/kg) in Stellenbosch, Western Cape, South Africa during the period of study. The environmental impacts were estimated using the values from previous studies reported in literature. The energy used for storage and processing activities and greenhouse gas (GHG) emission associated with fruit production were estimated using 6.1 MJ/kg and 0.48 CO_2 eq/kg, respectively [34]. The values were estimated for apples, which is a deciduous fruit like the pomegranate and which have similar packhouse processes. The water footprint was estimated with 910 m^3 ton^{-1} [35]. The nutritional impacts were calculated using values from [36] and [37]. Furthermore, cropland use was estimated by the size of the farm and the average yield produced.

2.3. Data Collection

The data collection protocol was consistent with the direct measurement method of the Food Loss and Waste Protocol (FLWP) [38]. For fruit loss data collection, a total of 251 bins (containing 1300–1500 fruit each) were put through the packhouse line and the number of waste bins (fruit loss) produced for each cultivar were recorded. Altogether, a total of over 351,400 individual fruit were assessed which comprised of 89, 84, and 78 bins of 'Acco', 'Herskawitz', and 'Wonderful' pomegranate, respectively. To determine the causes for the loss, fruit in 18 waste bins (6 per cultivar) were further examined. For each bin, a sample of 30 fruit was randomly selected each from the bottom, middle, and top and placed into ventilated cartons. Each fruit was visually assessed based on physical appearance (presence of rot, *Alternaria* disease, crack, injury, sunburn, blemish, insect damage), sorted, counted, and recorded according to each type of defect. In total, 1630 fruit (540 per cultivar) were examined to determine the quality defects causing fruit loss.

Data collection for each cultivar was done in three days, and six bins (n = 6) were assessed per cultivar ('Acco', 'Hershkawitz', and 'Wonderful'). The waste bins were labelled and two bins were assessed per day. It is important, however, to mention that pomegranate fruit losses at the packhouse level are not necessarily cultivar dependent, rather they originate from direct (primary sources) and indirect (secondary sources) [39]. Nonetheless, it was important to categorise fruit defects by cultivar for ease of data collection and comparison with historical packhouse data.

2.4. Historical Packhouse Data

Historical data on pomegranate postharvest fruit losses collected by quality control staff at the case study packhouse for the two years where data were available (2016 and 2019) were obtained as secondary data. These data are presented and discussed in comparison with the results obtained in the present study.

2.5. Statistical Analysis

Microsoft Excel 2013 (Microsoft Corporation) was used to collate the data collected. In order to find the trend of variation between cultivars and fruit defects and to consider their correlation, data were investigated according to principal component analysis (PCA) using XLSTAT software Version 2012.4.01 (Addinsoft, Paris). The mean value ± standard error of fruit defects was also presented and where there was a statistical significance difference ($p < 0.05$), analysis of variance (ANOVA) was performed using Statistica Version 13.5.0 to evaluate the differences between cultivars and fruit defects. Significant differences between means were separated using Duncan's multiple range test.

3. Results

3.1. Magnitude of Fruit Losses and Waste

The magnitude of pomegranate fruit losses at the packhouse was measured by the proportion of bins of discarded fruit to the number of bins initially put in the fruit processing line. Loss quantification involved a total of 251 bins of fruit put into the processing line from the three cultivars studied, of which 18 bins were discarded for failing to meet the minimum market required standard. The total lost fruit among the three cultivars ranged from 6.74 to 7.69% (Table 1). 'Acco' produced the least lost fruit as 89 fruit bins put in the processing line produced 6 bins of discarded fruit, while 84 fruit bins of 'Hershkawitz' produced 6 bins of discarded fruit. Lastly, 'Wonderful' produced the highest amount of lost fruit as 78 bins of fruit put in the processing line produced 6 bins of discarded fruit.

Table 1. Amount and percentages of each pomegranate cultivar fruit lost (discarded) based on the amount of fruit put through the packhouse line.

Pomegranate Cultivar	Fruit Put through the Processing Line (Bins)	Discarded Fruit (Bins)	Loss (%)
'Acco'	89.00	6.00	6.74
'Hershkawitz'	84.00	6.00	7.14
'Wonderful'	78.00	6.00	7.69
Total	251.00	18.00	21.5
Mean	83.60	6.00	7.16

Estimates of pomegranate fruit losses at the packhouse level in South Africa have been reported in recent years by the Pomegranate Association of South Africa (POMASA). In 2017, POMASA reported 11% loss in 'Wonderful', 13% loss in 'Hershkawitz', and 11% loss in 'Acco' [40]. In 2018, a 7% loss of 'Wonderful' was reported, an 8% loss in 'Hershkawitz', and a 9% loss in 'Acco' [18]. Additionally, in 2019, 9% of 'Wonderful' was reported as a loss, 25% loss in 'Hershkawitz', and 13% loss in 'Acco' [16]. Pomegranate fruit loss estimation at the packhouse is measured throughout the production season with fruit from multiple farmers with different preharvest and postharvest practices, which could affect the quality of fruit delivered to the packhouse and, hence, the amount of loss recorded. These factors account for the higher incidence of postharvest losses at the packhouse based on reported historical industry-wide data compared with the site-specific results obtained in the current study through a case study.

Bond [41] reported a 20% loss in carrots at the packinghouse level in Norway. The estimation was done using secondary data from experts in the carrot industry and surveys with semi-structured interviews with managers of packhouses. The study revealed that mechanical damage (harvesting technique at the farm) is a major source of loss at the packhouse since the superficial injuries during harvest open wounds for decay and disease infestation. A postharvest loss assessment of avocado, banana, guava, mango, papaya, and tomato was carried out among fruit growers and traders in north-western Ethiopia by Bantayehu et al. [42]. The results show that 18–28% of losses occurred during harvesting, storage, and transportation, while 18–25% of losses were reported at transportation and marketing levels. The major causes of loss are superficial injury, bruising, sunburn, handling technique, and physiological disorders, which are similar to the causes of pomegranate fruit loss in this present study. Semi-structured questionnaires and interviews were used for data collection in the study. Furthermore, a study in Nepal reported 35% loss in carrots [43]. Farmgate loss was estimated at 10%, 2% at a collection point, 5% at the wholesale market, and 18% at the retail level, and crack and splits were identified as the major cause of carrot loss [43]. Irrespective of the magnitude of loss reported in the studies, losses due to environmental stress and mechanical damage have remained dominant among the causes of fruit loss, which are similar to the results of this present study.

3.2. Causes of Packhouse Pomegranate Fruit Losses

The causes of packhouse pomegranate fruit losses were assessed based on the quality issues of why fruit were removed from the packhouse processing line as waste. These quality issues have contributory factors, and some are direct (primary source) while some are indirect (secondary source) [39]. The main indirect (secondary) cause of packhouse pomegranate fruit loss is the high market standard. South Africa exports about 76% of the total pomegranate production [16] and 61% of the total export goes to the European markets [17]. The trend of pomegranate marketing in Europe shows that South Africa faces strong competition with other countries of the Southern Hemisphere for the market share [1,16]. This competition is believed to have raised the market standard, which means that only premium quality fruit are processed for export at the packhouse. The implication of this is that pomegranate fruit are sorted again at the packhouse to ensure that only the best quality fruit are packed for sale. The 'good fruit' that are deemed not to meet the premium quality required in the export market are sold locally. The effect of this is that more fruit are lost or sold at a cheap price for juicing and other purposes. Additionally, handling at the packhouse is another source of loss categorised as a direct (primary) source of loss. Losses due to handling manifested mainly as fruit bruises and superficial injuries. However, the two major reasons for physical loss as identified in this study were sunburn and injury. Other reasons are *Alternaria*, bruises, cracks, being oversized, insect damage, rot, decay, blemishes, and malformation.

3.2.1. Environmental Stress (Sunburn, Cracks, and Splits)

Sunburn

In the three cultivars assessed, sunburn was recorded as the highest cause of loss. Losses due to sunburn at the packhouse originated from the farm where pomegranate fruit were exposed to direct sunlight, which causes discolouration of the rind of the affected fruit, hence downgrading the fruit quality [44]. This shows the effect of high temperature on the quality of pomegranate fruit. After sorting for premium quality fruit at the packhouse, sunburn accounted for 28.70% and 29.8% of the discarded fruit in 'Acco' and 'Hershkawitz', respectively (Table 2). The highest fruit loss incidence was in 'Wonderful', where it contributed to 34.81% of losses. Sunburn showed a positive relationship with oversized fruit in the correlation analysis result (Table 3). This relationship is the only positive relationship result in the analysis, which indicates that more oversized fruit with sunburn were deemed fit for export at the farm level but could not meet the minimum market standard according to the evaluation of the packhouse. The market standard in Europe and the Middle East does not allow fruit with noticeable sunburn, which means that such fruit are sold at a low price locally, mainly for juicing.

Table 2. Percentage fruit loss of three pomegranate cultivars due to different defects at packhouse.

	Cultivar		
Fruit Defect	**'Acco' (Loss %)**	**'Hershkawitz' (Loss %)**	**'Wonderful' (Loss %)**
Alternaria	4.30	3.10	2.96
Bruise	13.33	12.80	10.94
Injury	23.33	23.70	19.07
Sunburn	28.70	29.80	34.81
Crack	18.70	18.34	17.96
Insect damage	3.90	2.20	2.77
Crown rot	2.22	2.96	1.67
Decay	2.22	1.90	2.22
Blemish	3.30	3.30	3.70
Misshapen	0.00	1.90	1.66
Oversized	0.00	0.00	2.24

Table 3. Pearson correlation coefficient matrix between defects on three pomegranate cultivars ('Acco', 'Hershkawitz', and 'Wonderful').

Defects	*Alternaria*	Oversized	Bruise	Injury	Sunburn	Crack	Insect Damage	Crown Rot	Decay	Blemish	Misshapen
Alternaria	1										
Oversized	−0.267	1									
Bruise	−0.248	−0.179	1								
Injury	0.157	**−0.356**	−0.170	1							
Sunburn	−0.020	**0.376**	−0.179	−0.267	1						
Crack	−0.246	−0.112	−0.230	−0.184	**−0.402**	1					
Insect damage	0.116	−0.133	−0.157	−0.219	−0.208	0.157	1				
Crown rot	0.067	−0.218	−0.093	0.042	−0.181	−0.063	0.003	1			
Decay	−0.088	0.108	−0.209	0.041	−0.208	−0.076	0.055	−0.011	1		
Blemish	**−0.362**	0.131	−0.100	−0.112	−0.094	0.096	−0.124	−0.160	−0.008	1	
Misshapen	−0.088	0.201	−0.007	**−0.327**	−0.007	0.126	−0.257	−0.130	0.002	0.011	1

Values in bold are significant at $p < 0.05$.

Temperatures exceeding 35 °C and low relative humidity at the farm level contribute to a higher incidence of sunburn [45] and because 'Wonderful' pomegranate produces bigger fruit with a larger surface area and is a late cultivar in South Africa, this results in fruit hanging on the tree much longer before harvest. With most of the fruit exposed to direct sunlight outside the tree canopy, the incidence of sunburn is exacerbated. This combination of factors makes 'Wonderful' pomegranate fruit more susceptible to sunburn than 'Acco' and 'Hershkawitz'.

Cracks and Splits

The results show that the amount of fruit affected by cracks and splits in the three cultivars studied are similar as they ranked third in the causes of loss in the cultivars. The highest incidence was in 'Acco', where they accounted for 18.70% of losses (Table 2). For 'Hershkawitz', cracks and splits contributed to 18.34%, while in 'Wonderful', they accounted for 17.96% of losses. Cracks and splits had a negative relationship with sunburn according to the correlation analysis result (Table 3). This shows the impact of fruit sorting at the farm level; otherwise, it is reasonable to believe that higher sunburn would result in more cracks and splits due to the hardening of fruit rinds due to direct sunlight, which aids cracking when the moisture content fluctuates. Like sunburn, pomegranate cracks and splits as observed at the packhouse mostly originated from the farm and were a result of environmental stress, specifically soil moisture imbalances [46,47] as pomegranate fruit are highly sensitive to variation in the soil moisture content [48]. Therefore, fruit with cracks and splits at the packhouse are due to either oversight by the farm fruit sorters or the assumption that the fruit could meet the minimum market standard.

Cracks and splits create an open wound that enhances moisture loss and disease infestation, which lowers the quality of the affected fruit [49]. Fruit discarded from the packhouse due to cracks and splits were sold locally for industrial use.

3.2.2. Mechanical and Physical Damage (Superficial Injuries, Bruise Damage, and Blemishes)

Superficial Injuries

Superficial injuries were the second highest cause of pomegranate fruit loss at the packhouse after sunburn. Injuries constituted 23.33% of the total loss in 'Acco' (Table 2). For 'Hershkawitz', injury contributed 23.70% of the loss, which is the highest incidence of injury recorded among the three studied cultivars. 'Wonderful' recorded the least amount of injury with 19.07% of losses in the cultivar. Superficial injuries showed a negative relationship with oversized fruit in the correlation matrix (Table 3). This indicates that a higher incidence of injury was due to handling and not fruit sizes. Some of the superficial injuries observed were cases of opening fruit with a suspicion of internal disease by packhouse fruit sorters with false results. Furthermore, losses due to injuries originating from preharvest and handling technique at the farm level were observed. Injuries in this category were deemed insignificant at the farm level, but the affected fruit failed to meet market standards by the packhouse. Pomegranate fruit were only stored for a few days (when necessary) at the packhouse before they were processed; therefore, chilling injuries were not observed.

Bruise Damage

The results show that bruise damage is the fourth cause of loss in the three pomegranate cultivars assessed. 'Acco' recorded the highest incidence of bruise damage, which accounted for 13.33% of losses in the cultivar (Table 2). Bruise damage contributed to 12.80% of the losses recorded for 'Hershkawitz' and 10.94% of the losses in 'Wonderful'. Bruise damage showed no significant relationship with any other defect in the correlation analysis result (Table 3), which suggests that bruise damage at the packhouse is solely a function of mechanical damage during transportation and handling at the packhouse.

Like an injury, a bruise is caused by mechanical damage as a result of impact during harvesting, transportation, and handling [50]. Most of the bruises observed were believed to occur during transportation to the packhouse and packhouse handling. Many farm roads are rough, thereby causing vibration and compression of the fruit during transportation, which results in bruising damage [50,51]. Moreover, vibration and impact occur during fruit unloading at the packhouse and conveyance to the processing line. These assumptions were made because the affected areas of the fruit were already brownish in colour and soft, illustrating that the bruising was not an immediate occurrence. However, there were cases where the affected fruit were discarded during packaging with no visible discolouration of the rind but with softness in the affected parts. Bruised fruit do not meet either the export or the local market standards, and therefore, are sold at a low price for industrial use.

Blemish

Blemish is one of the least frequent causes of loss in the three pomegranate cultivars studied. For 'Acco', it ranked seventh out of eight in the causes of loss and accounted for 3.30% of losses. It ranked fifth in 'Hershkawitz' and contributed to 3.30% of fruit loss. The highest occurrence of blemish was recorded in 'Wonderful' with 3.70% of losses. The presence of fruit with blemish at the packhouse is usually the result of oversight from the on-farm fruit sorters as they are unlikely to be caused by packhouse handling operations. Blemish is mostly a result of mechanical damage during and after pruning before pomegranate fruit are picked. Again, sharp tree branches scratch fruit when thrown against them by the wind, leaving blemish marks on the affected fruit. Blemish is a strong factor in determining pomegranate fruit quality both for export and local market because external attractiveness of pomegranate fruit depends strongly on a blemish-free appearance [52].

3.2.3. Biological Damage (Insect Damage)

Insect Damage

The results show that insect damage contribution to pomegranate fruit losses at the packhouse was low. The highest incidence of insect damage was in 'Acco', where it ranked sixth in the causes of loss and accounted for 3.90% of losses (Table 2). The lowest incidence was in 'Hershkawitz' with 2.20% of losses and ranked eighth in the causes of loss. For 'Wonderful', it accounted for 2.77% of losses. Insect damage had no significant relationship with other defects assessed in the correlation analysis (Table 3). This indicates that insect damage in this present study occurred independently of other defects and that it was not as a result of packhouse operation. It could also mean that a significant amount of fruit damaged by insects were discarded at the farm level.

Insect damages downgrade the quality of pomegranate fruit since a small portion of the fruit is consumed, which results in a partial loss of the affected fruit and in making them not meet market standard. The affected fruit were discarded from the processing line and sold at a low price since part of the fruit could still be used for other purposes such as the manufacturing of dye and animal feed.

3.2.4. Microbial and Pathological Spoilage (Decay and Rots, *Alternaria*, Crown Rot)

Decay and Rots

Decay and rots are one of the least frequent causes of pomegranate fruit loss among the three cultivars assessed at the packhouse. For 'Acco', it accounted for 2.22% of losses (Table 2). They contributed to 1.90% of losses in 'Hershkawitz' and 2.22% in 'Wonderful'. Decay and rot had no significant relationship with other defects in the correlation analysis (Table 3). This indicates that decay at the packhouse, in this present study, was not a result of packhouse operation (handling). Therefore, the decayed fruit were because of a sorting oversight at the farm level. Decay and rot are a result of microbial pathogens that break down the rind of the affected fruit, which results in partial or total decay and rot [20]. Decayed fruit do not meet market standard and are often buried or composted.

Alternaria

Alternaria disease varied among the three studied cultivars at the packhouse. However, its contribution to total fruit loss was low. The highest incidence of *Alternaria* was in 'Acco', where it contributed to 4.30% of losses (Table 2). For 'Hershkawitz', *Alternaria* accounted for 3.10% of loss and ranked sixth in the causes of loss in the cultivar, and contributed 2.96% of loss in 'Wonderful'. *Alternaria* disease occurs at the farm and fruit discarded at the packhouse due to the disease were due to a sorting oversight at the farm because often, it is difficult to determine infected fruit physically.

Alternaria is a pomegranate fruit disease caused by the *Alternaria alternata* pathogen. The disease causes fruit to decay partially or totally from the inside. In contrast, the rind of the affected fruit appears healthy [19]. The affected fruit are light in weight, which makes them float during chlorine baths at the packhouse processing line. *Alternaria*-affected pomegranate fruit are intensely reddish in colour compared to an *Alternaria*-free fruit. These fruits are often buried or composted.

Crown Rot

Crown rot accounted for a low amount of pomegranate fruit loss at the packhouse. It contributed to 2.22% of losses in 'Acco' (Table 2). The highest occurrence of crown rot was in 'Hershkawitz', where it accounted for 2.96% of loss and ranked seventh in the causes of loss for the cultivar. For 'Wonderful', it ranked tenth in the causes of loss and accounted for 1.67% of losses. Crown rot showed no significant relationship with other defects in the correlation analysis (Table 3), which suggests that it occurred for reasons outside of the packhouse. Like *Alternaria*, crown rot is a farm disease and did not originate at the packhouse, rather, it was found due to a sorting oversight at the farm.

Crown rot is caused by *Coniella granati*, a fungi pathogen [19], which mostly affects pomegranate fruit on the farm. The rind of the affected fruit shows the presence of pycnidia with rotten crown [19]. Fruit affected by crown rot were discarded for not meeting the market standard, and as such, were sold at a cheap price for industrial products such as ink and dye.

3.2.5. Irregular Fruit Size and Shape (Oversized and Misshapen)

Oversized

Oversized fruit were only observed among the 'Wonderful' cultivar and in a very small quantity. Therefore, oversized fruit contributed little to overall pomegranate fruit loss in the cultivar. Oversized fruit accounted for 2.24% of loss (Table 2). The oversized fruit were not able to fit comfortably into the 3.8 kg equivalent carton used for pomegranate fruit packaging, and therefore, were sorted to be sold and used for other purposes such as juicing.

Misshapen

Pomegranate fruit discarded for being misshapen were very few and contributed least to the causes of loss. Such fruit were found only in 'Hershkawitz' and 'Wonderful'.

For 'Hershkawitz', it contributed to 1.90% of loss, and in 'Wonderful', 1.66% (Table 2). The misshapen fruit were good fruit with irregular shapes, hence, they did not appear appealing for the shelves but could be used for producing juice, jam, and dye.

3.3. *Comparative Analysis of Pomegranate Fruit Based on Defects*

Fruit were discarded from the processing line for not meeting market standard due to bruising and injury (during handling), and other defects such as sunburn and microbial and pathological diseases that originated from the farm. Although packhouse defects are not considered cultivar-dependent, this study evaluated the relationship between pomegranate fruit defects and the cultivars using principal component analysis (PCA). The result was observed in biplot axes, which shows a relationship by the clustering of active variables (defects), in the red colour, around active observations (cultivars), in the blue colour (Figure 1). The result revealed that oversized and misshapen fruit were most common amongst the 'Wonderful', as evidenced by their clustering around 'Wonderful'. At the same time, insect damage and *Alternaria* were predominant in 'Acco'. Decay and crown rot were primarily associated with 'Herskawitz'. Bruise and injury, which are mainly due to fruit handling, were observed to affect the three cultivars relatively equally. Environmental stress factors (sunburn and cracks) were also found to affect the three cultivars in a similar proportion. A dendrogram cluster analysis was done to evaluate whether different packhouse management practices would be advisable for the handling of each cultivar (Figure 2). The result suggests that implementing different packhouse management practices is not necessary for each cultivar as the three cultivars clustered around each other in cluster 2 and 3, which supports the fact that packhouse fruit loss is not cultivar dependent, rather due to postharvest handling practices and preharvest factors which caused some of the defects ab initio. Cluster 1 consists only of 'Wonderful', and this could be attributed to misshapen and oversized fruit, which were majorly associated with the cultivar.

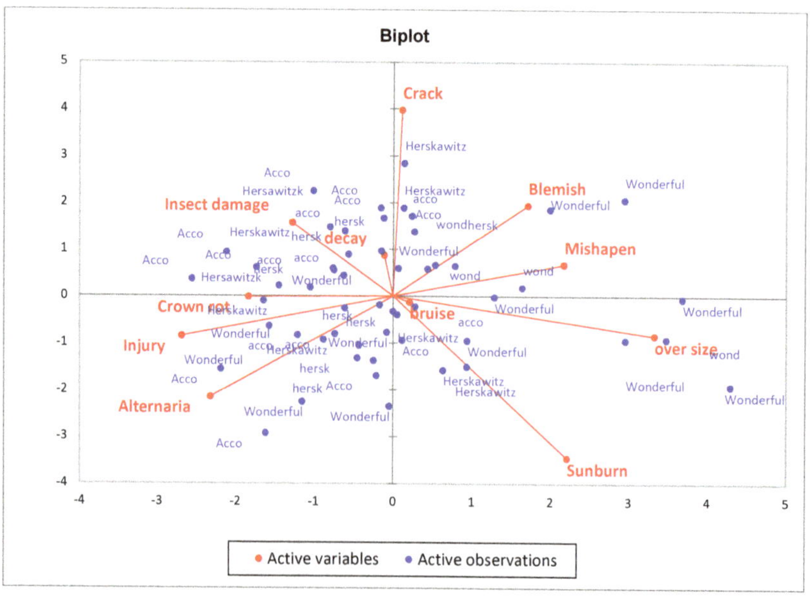

Figure 1. Observation chart showing fruit defects according to cultivars.

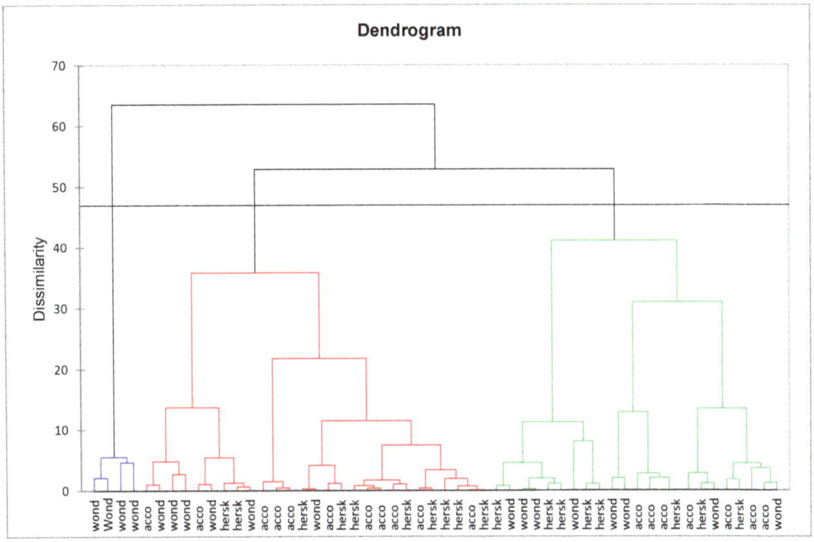

Figure 2. Dendrogram of cluster analysis of three pomegranate cultivars studied based on defects. Key: wond = 'Wonderful', acco = 'Acco', hersk = 'Hershkawitz'.

The analysis of variance (ANOVA) was performed to evaluate differences in how the defects affect cultivars, as presented in Table 4. The effects of defects on 'Acco' and 'Hershkawitz' were similar but different in 'Wonderful' except for sunburn and superficial injury. The defects originated from sources such as environmental stress, mechanical and physical damage, biological damage, microbial and pathological spoilage, and lastly, irregular fruit size and shape (Table 4). The results show that environmental stress was the major cause of pomegranate fruit losses at the packhouse. However, it is important to note that the environmental factors originated from the farms and the affected fruit were discarded at the packhouse as they did not meet the required market standard. Environmental stress accounted for the highest incidence of loss, with 49.44% of the total losses. Mechanical and physical damage also caused significant loss of fruit, accounting for 37.84% of total fruit losses. The biological damage factor was only insect damage, which contributed 2.96% of losses while irregular fruit size and shape contributed least to losses with 1.92% and were mostly in 'Wonderful'. Lastly, microbial and pathological spoilage accounted for 7.84% of total losses.

Table 4. Comparison between cultivars and fruit defects contributing to postharvest loss in the case study packhouse.

	Cultivar							
Defects	**Acco (Mean)**	**Acco (Total)**	**Hershkawitz (Mean)**	**Hershkawitz (Total)**	**Wonderful (Mean)**	**Wonderful (Total)**	**Total**	**Loss (%)**
Biological								
Insect damage (mean)	3.50 ± 0.72 [e,*]	21	2.00 ± 0.68 [e,f]	12	2.50 ± 0.43 [d]	15		
Total		21		12		15	48	2.96
Irregular fruit size and shape								
Misshapen	0.00 ± 0.00 [f]	0	1.67 ± 0.42 [e,f]	10	1.50 ± 0.43 [d]	9		
Oversized	0.00 ± 0.00 [f]	0	0.00 ± 0.00 [f]	0	2.00 ± 0.63 [d]	12		
Total		0		10		21	31	1.92
Mechanical damage								
Bruise damage	12.00 ± 1.61 [d]	72	11.50 ± 0.43 [d]	69	9.83 ± 1.08 [c]	59		
Superficial injuries	21.00 ± 0.73 [b]	126	21.33 ± 0.80 [b]	128	17.17 ± 0.54 [b]	103		
Blemish	3.00 ± 0.52 [e]	18	3.00 ± 0.37 [e]	18	3.33 ± 0.42 [d]	20		
Total		216		215		182	613	37.84
Environmental stress								
Sunburn	25.83 ± 0.87 [a]	155	26.83 ± 1.47 [a]	161	31.33 ± 0.61 [a]	188		
Cracks and splits	16.83 ± 1.08 [c]	101	16.50 ± 1.28 [c]	99	16.17 ± 1.14 [b]	97		
Total		256		260		285	801	49.44
Microbial and pathological								
Alternaria	3.83 ± 0.83 [e]	23	2.83 ± 0.70 [e]	17	2.67 ± 0.33 [d]	16		
Crown rot	2.00 ± 0.26 [e,f]	12	2.67 ± 0.61 [e]	16	1.50 ± 0.50 [d]	9		
Decay and rots	2.00 ± 0.37 [e,f]	12	1.67 ± 0.21 [e,f]	10	2.00 ± 0.63 [d]	12		
Total		47		43		37	127	7.84

* Mean values in the same row followed by different letters (a–f) indicate significant differences ($p < 0.05$).

4. Discussion

4.1. Historical Packhouse Data on Pomegranate Fruit Losses at Case Study Packhouse in Wellington, Western Cape, South Africa

Historical fruit loss data for 2016 and 2019 at the case study packhouse were analysed in comparison with the results of this present study and presented in Figure 3. The result suggests that marketing standard is a major source of fruit loss at the packhouse. This means that some fruit deemed suitable for marketing (export and local) at the farm level do not meet the packhouse marketing standard as a result of defects originating from the farm. This is evident in the contribution of sunburn and cracks to fruit losses as compared to bruise and injury, which are believed to be because of transportation and handling at the packhouse level. Furthermore, blemish, which also originates from the farm, was found to account for a significant amount of fruit loss at the packhouse according to both the packhouse historical data and the result obtained from the present study.

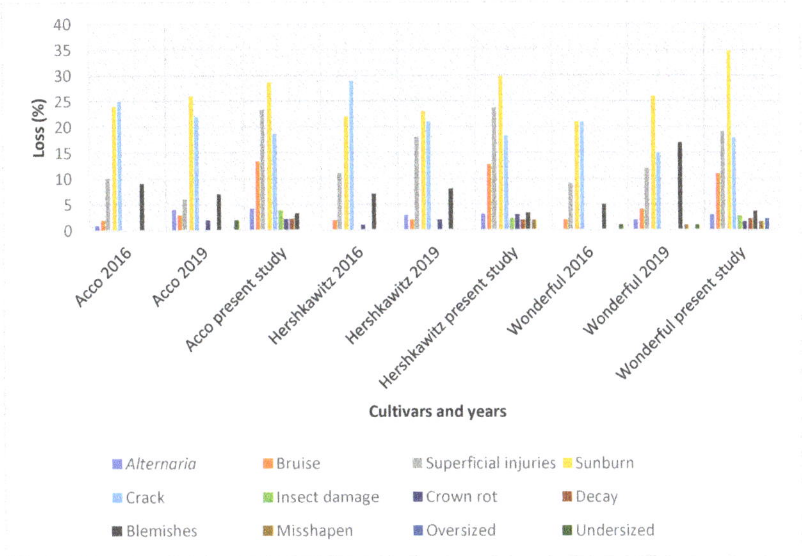

Figure 3. Comparison of historical packhouse pomegranate fruit defect data (2016 and 2019) and the present study.

4.2. Economic, Environmental, and Resource Impacts

The impacts of pomegranate fruit loss estimated in this study are based on the magnitude of incidence of pomegranate fruit loss at the case study packhouse in Wellington, Western Cape Province and retail price in South Africa. This is to reveal the potential production inputs and resources that are wasted in producing the pomegranate fruit that are lost. For example, the energy used for the production of wasted food could be used for another productive purpose such as cold storage to preserve food. Typically, the amount of packhouse fruit loss at the national and global level might be different depending on a range of factors including production practices, postharvest handling, and the market standard at the importing markets. The estimations are particularly important to raise awareness on the importance of reducing fruit losses at the packhouse level given several sustainability challenges that the world is facing, which require prudent use of resources today to create a future with sufficient material and natural resources [53].

The retail price of pomegranate fruit at the supermarket means that ZAR 88.99 (USD 5.26) is lost per 1 kg of lost pomegranate fruit in South Africa. Based on the annual average loss of 7.16% at the case study packhouse (Table 5), which translates to 328.79 tonnes, the monetary loss of the total annual production was estimated at ZAR 29.5 million (USD 1,754,984). During the production of pomegranate fruit, greenhouse gases(GHGs) are emitted into the atmosphere. Based on the findings of this study, the pomegranate losses at the packhouse level were estimated to emit about 157,819 CO_2 eq. To sink this amount of CO_2 eq would require planting about 4 million trees at 0.039 metric ton CO_2 per tree planted [54]. Furthermore, an estimated 2,005,619 MJ of energy and 299,198.9 m^3 of water were wasted in production. This amount of wasted water could meet the daily water requirement of up to 109,896 persons in a year at 0.05 m^3 utilised per person per day [55]. Again, the production of the lost fruit could take up to 8.54 ha of land, that could have otherwise been used to provide public utilities such as a shopping complex.

Table 5. Summary of the magnitude of pomegranate fruit losses and impacts at the packhouse, South Africa, and global levels.

Factors	Case Study Packhouse	South Africa	Global
Production volume (tonnes) *	4592.00	32,572.11	3139×10^3
Average loss (%)	7.16	7.16	7.16
Retail price (ZAR/kg) [a]	89.99	89.99	89.99
Estimated physical and economic losses			
Physical loss (tonnes)	328.79	2,332.16	224,792.50
Monetary loss (ZAR)	29×10^6	209×10^6	$20{,}229 \times 10^6$
Environmental impacts			
Estimated GHG emission (CO_2 eq) [b]	157×10^3	1×10^6	107×10^6
Estimated energy used (MJ) [c]	2×10^6	14×10^6	1371×10^6
Resource impact			
Water footprint (m^3) [d]	299×10^3	2122×10^3	$204{,}561 \times 10^3$
Equivalent land used to produce lost fruit (ha)	8.54	60.58	5838.77

* Production statistics is estimated from Sonlia packhouse [56]. [a] Supermarket retail price in Stellenbosch, Western Cape, South Africa. [b,c] Impacts per unit fruit produced estimated from [34]. [d] Impact per unit fruit produced estimated from [35].

The economic and environmental impacts of pomegranate fruit losses at packhouse were also estimated at the national (South Africa) level. Losses at the national level were estimated at 2332.16 tonnes (Table 5), which translates to an estimated ZAR 209.87 million (USD 12.64 million) annual revenue loss. Losses at the national level were found to emit about 1.11 million CO_2 eq. To sink this amount of CO_2 eq would require planting at least 28 million trees at 0.039 metric ton CO_2 per tree planted [54]. Furthermore, about 14.22 million MJ of energy and 2.12 million m^3 of water were wasted to grow the lost fruit. The wasted water could meet the daily water requirement of about 116,289 people for a year at 0.05 m^3 consumed per person per day [55]. Lastly, the land used to produce the lost fruit was estimated at 60.58 ha of land.

Furthermore, the economic and environmental impacts of pomegranate fruit losses were estimated at the global level using the incidence of losses and retail price in South Africa. This assumes a 7.16% loss of total fruit conveyed to the packhouse for processing globally, which was estimated at 224,792 tonnes (Table 5) and a retail price of ZAR 88.99/kg (USD 5.26/kg). The revenue loss due to the lost fruit was estimated at ZAR 20.22 billion (USD 1.2 billion). Based on the estimation, about 107.90 million CO_2 eq were emitted annually due to losses of pomegranate fruit. To sink this amount of CO_2 eq would require planting at least 2.7 billion trees at 0.039 metric ton CO_2 per tree planted [54]. Additionally, about 1.37 billion MJ of energy and 204.56 million m^3 of freshwater were wasted. The wasted water could meet the daily water requirement of about 11.2 million people for a year at 0.05 m^3 utilised per person per day [55]. Lastly, about 5838.77 ha of land was used to produce the lost fruit. Postharvest losses of pomegranate fruit mean a significant loss of revenue and resources that could have otherwise been put to beneficial use.

4.3. Nutritional Impacts

The loss of pomegranate fruit contributes to food and nutritional insecurity in South Africa due to a huge loss of essential nutrients in the lost pomegranate fruit. Some of the nutrients lost due to postharvest losses at the case study packhouse in Wellington, Western Cape Province of South Africa during the 2020 season are presented in Table 6. The nutritional impacts of fruit and vegetable cannot be over-emphasised, especially given the effect of the COVID-19 pandemic on the livelihood of individuals and their ability to afford healthy and nutritious food. Based on the annual loss of pomegranate fruit during operations at the case study packhouse, the lost content of sodium, fibre, carbohydrate, iron, and ascorbic acid in fruit were estimated to meet the daily recommended nutrition intake of 1, 7, 25, 5, and 66 people, respectively.

Table 6. Selected nutritional impacts of pomegranate fruit losses at the case study packhouse, Wellington, in the Western Cape Province of South Africa.

	Case Study Packhouse		National (South Africa)		Global	
Nutrition factor	Amount lost ($mg100^{-1}$ g) *	Nutritional loss (per capita/day) **	Amount lost ($mg100^{-1}$ g) *	Nutritional loss (per capita/day) **	Amount lost ($mg100^{-1}$ g) *	Nutritional loss (per capita/day) **
Fibre	164.39 ##	7.00	1166.08 ##	47.00	112×10^3 ##	4496.00
Carbohydrate	3255.02 ##	25.00	23,088.38 ##	178.00	222×10^4 ##	17,119.00
Protein	460.30 ##	10.00	3265.02 ##	71.00	314×10^3 ##	6842.00
Iron	98.64	5.00	699.65	39.00	67×10^3	3747.00
Ascorbic acid	4931.85	66.00	34,982.40	466.00	337×10^4	44,959.00
Calcium	9863.70	10.00	69,964.80	70.00	674×10^4	6744.00
Magnesium	3945.48	13.00	27,985.92	90.00	269×10^4	8702.00
Sodium	1315.16	1.00	9328.64	5.00	899×10^3	450.00
Potassium	56,223.09	12.00	398,799.40	85.00	384×10^5	8179.00

* Amount lost is based on [32]. ** Nutritional loss is based on [31]. ## Amount lost is estimated in $g100^{-1}$ g.

The nutritional impacts of pomegranate fruit losses were also estimated at the national (South Africa) level using the incidence of losses at the case study packhouse, in the Western Cape Province of South Africa (Table 6). Based on the annual losses of pomegranate fruit at the packhouse level, the estimate at the national level suggests that the lost content of sodium, fibre, calcium, magnesium, and ascorbic acid in fruit could meet the daily recommended intake of 5, 47, 70, 90, and 466 people, respectively.

The estimation of postharvest nutritional losses of pomegranate fruit at the global level showed a huge loss of essential nutrients that could benefit people in a period where micro and macronutrient deficiency affects not less than a third of the world population and negatively impacts the quality of life [57]. Based on the annual incidence of losses in South Africa, the selected nutrient loss globally due to pomegranate losses at the packhouse was estimated (Table 6). The lost content of sodium, fibre, protein, potassium and ascorbic acid in fruit could meet the daily recommended nutrition intake of 450, 4496, 6842, 8179 and 44,959 people respectively. The findings revealed that postharvest losses of pomegranate fruit at the packhouse level also contribute to global food and nutrition insecurity.

4.4. Possible Solutions to Overcome and Limit Fruit Loss at the Packhouse

This study identified quality issues that lead to the downgrading of a significant proportion of the fruit processed at the packhouse. The quality issues are categorised into indirect (secondary) and direct (primary) causes of loss [39]. Indirect sources of loss are mainly due to high market quality standard at the importing markets. At the case study packhouse, pomegranate fruit that did not meet market quality standard were majorly due to the loss of aesthetic and physical appeal because of defects and damages leading to downgrading. Fruit losses due to high market quality standards could be classified as unavoidable loss [58]. This is because market quality standards are determined by market specifications on produce quality attributes and economic factors that are beyond the control of the packhouse operation. Under this situation, the application of the best available postharvest technologies at the packhouse cannot prevent such losses due to products that fall short of market standards.

Fruit losses due to direct (primary) causes at the case study packhouse include losses due to postharvest handling of fruit during transportation from storage to the processing line, sorting and grading; these are manifested as superficial injuries, cuts, and bruises [50]. Since poor postharvest handling practices are a major cause of fruit loss, possible solutions to reduce loss must be practical and technologically driven [59,60]. Fruit quality improvement can be achieved by local investment in technological innovations through research to improve knowledge in life cycle assessment, processing, and handling of pomegranate fruit at the packhouse [59,61,62]. The conventional manual sorting technique used at the packhouse is subjective and often leads to damaging wholesome fruit because fruit sorters most times are unable to make distinction between internally damaged fruit and a good fruit. Hence, possible technological improvements in the packhouse line such as non-destructive sorting techniques using remote sensing along the processing line would limit basic sorting errors leading to cutting fruit open to ascertain the presence of internal diseases. In addition to technological innovation, the reduction of fruit loss will require the continuous training of packhouse staff on fruit handling, especially the fruit sorters and forklift drivers. This is important because reckless transportation from temporal cold storage to the packhouse processing line causes bruises, which lead to fruit loss. This could be limited by educating forklift drivers about the impact of vibration and compression on pomegranate fruit.

5. Conclusions

This study found that pomegranate fruit loss at the case study packhouse in Wellington, Western Cape Province of South Africa ranged between 6.74 to 7.69%. This translates to 328.79 tonnes of pomegranate fruit removed from the packhouse processing line per production season. This amount of fruit is removed from the value chain for not meeting the minimum market standard and are sold at a low price for juicing and as raw material for dye and ink production. The major direct cause of pomegranate fruit loss at the packhouse, as identified in this study is handling (bruise and injuries). Environmental stress (sunburn and cracks) and microbial and pathological diseases were also contributors to loss. It is interesting to note that the result of the causes of loss in this study is similar to the historical packhouse report as analysed.

The result of the magnitude of losses shows that the incidence of loss was lowest in 'Acco' with 6.74% of losses. The amount of loss in 'Herskawitz' and 'Wonderful' were similar with 7.14% and 7.69%, respectively. Market standard (especially the export market) is greatly influential on the amount of losses recorded at the packhouse. This is because most of the produce are exported to Europe and the Middle East, where only premium quality fruit are accepted. This means that fruit deemed marketable at the farm level may be discarded at the packhouse resulting in loss.

Packhouse fruit losses have a huge economic, environmental, resource, and nutritional impacts as exemplified in this study. The economic impact reflects the loss of revenue by farmers and other actors along the value chain. Environmental and resource impacts are evident in the unsustainable use of resources to produce lost and wasted fruit, and the nutritional impact results in food insecurity due to the wasted nutrients that would have otherwise benefitted people. Considering the various impacts of postharvest losses at the packhouse level, postharvest losses and waste reduction is a sustainable means of ensuring food and nutritional security. Furthermore, reducing postharvest losses and waste would help mitigate the effects of global warming and increase revenue for the food value chain actors.

Author Contributions: Conceptualisation, U.L.O.; methodology, I.K.O.; software, O.A.F.; validation, U.L.O. and O.A.F.; formal analysis, O.A.F.; investigation, I.K.O.; resources, U.L.O.; data curation, I.K.O.; writing—original draft preparation, I.K.O.; writing—review and editing, O.A.F.; visualisation, I.K.O.; supervision, U.L.O., O.A.F.; project administration, U.L.O.; funding acquisition, U.L.O. All authors have read and agreed to the published version of the manuscript.

Funding: This work is based on the research supported by the National Research Foundation of South Africa (Grant Numbers: 64813). The opinions, findings, and conclusions or recommendations expressed are those of the author(s) alone, and the NRF accepts no liability whatsoever in this regard. The APC was funded by the NRF (Grant number 64813).

Institutional Review Board Statement: Not applicable.

Informed Consent Statement: Not applicable.

Data Availability Statement: The data are not publicly available to protect the privacy of the case study packhouse.

Conflicts of Interest: The authors declare no conflict of interest.

References

1. Kahramanoglu, I. Trends in pomegranate sector: Production, postharvest handling and marketing. *Int. J. Agric. For. Life Sci.* **2019**, *3*, 239–246.
2. Kahramanoglu, I.; Usanmaz, S. *Pomegranate Production and Marketing*, 1st ed.; CRC Press: Boca Raton, FL, USA, 2016.
3. Middha, S.K.; Usha, T.; Pande, V. A review on antihyperglycemic and antihepatoprotective activity of eco-friendly Punica granatum peel waste. *Evid. Based Complement. Altern. Med.* **2013**, *656172*, 1–10. [CrossRef]
4. Pareek, S.; Valero, D.; Serrano, M. Postharvest biology and technology of pomegranate. *J. Sci. Food Agric.* **2015**, *95*, 2360–2379. [CrossRef]
5. Fawole, O.A.; Opara, U.L. Seasonal variation in chemical composition, aroma volatiles and antioxidant capacity of pomegranate during fruit development. *Afri. J. Biotechnol.* **2013**, *12*, 4006–4019.
6. Al-Said, F.A.; Opara, L.A.; Al-Yahyai, R.A. Physico-chemical and textural quality attributes of pomegranate cultivars (Punica granatum L.) grown in the Sultanate of Oman. *J. Food Eng.* **2009**, *90*, 129–134. [CrossRef]
7. de Oliveira, F.L.; Arruda, T.Y.P.; da Silva Lima, R.; Casarotti, S.N.; Morzelle, M.C. Pomegranate as a natural source of phenolic antioxidants: A review. *J. Food Bioact.* **2020**, *9*, 10–22. [CrossRef]
8. Faria, A.; Calhau, C. The bioactivity of pomegranate: Impact on health and disease. *Crit. Rev. Food Sci. Nutr.* **2011**, *51*, 626–634. [CrossRef]
9. Mphahlele, R.R.; Fawole, O.A.; Opara, U.L. Influence of packaging system and long term storage on physiological attributes, biochemical quality, volatile composition and antioxidant properties of pomegranate fruit. *Sci. Hort.* **2016**, *211*, 140–151. [CrossRef]
10. Heber, D.; Schulman, R.N.; Seeram, N.P. (Eds.) *Pomegranates: Ancient Roots to Modern Medicine*; CRC Press: Boca Raton, FL, USA, 2006.
11. Kang, N.J.; Shin, S.H.; Lee, H.J.; Lee, K.W. Polyphenols as small molecular inhibitors of signaling cascades in carcinogenesis. *Pharmacol. Ther.* **2011**, *130*, 310–324. [CrossRef] [PubMed]
12. Katz, S.R.; Newman, R.A.; Lansky, E.P. Punica granatum: Heuristic treatment for diabetes mellitus. *J. Med. Food* **2007**, *10*, 213–217. [CrossRef]
13. Wang, P.; Zhang, Q.; Hou, H.; Liu, Z.; Wang, L.; Rasekhmagham, R.; Kord-Varkaneh, H.; Santos, H.O.; Yao, G. The effects of pomegranate supplementation on biomarkers of inflammation and endothelial dysfunction: A meta-analysis and systematic review. *Complement. Ther. Med.* **2020**, *49*, 1–8. [CrossRef]
14. Işık Özgüven, A.; Gültekin, U.; Gözlekçi, S.; Yilmaz, I.; Yilmaz, C.; Küçük, E.; Imrak, B.; Korkmaz, C. A review of the economics and the marketing of the pomegranate industry in turkey. *Acta Hortic.* **2015**, *1089*, 221–228. [CrossRef]
15. Rymon, D. Mapping features of the global pomegranate market. *Acta Hortic.* **2011**, *890*, 599–602. [CrossRef]
16. Pomegranate Producers Association of South Africa (POMASA). Pomegranate Industry Overview: Statistics and Information. 2019. Available online: https://www.sapomegranate.co.za/statistics-and-information/pomegranate-industry-overview/ (accessed on 12 March 2020).
17. Pomegranate Producers Association of South Africa (POMASA). Pomegranate Industry Overview: Statistics and Information. 2018. Available online: https://www.sapomegranate.co.za/statistics-and-information/pomegranate-industry-overview/ (accessed on 12 March 2020).
18. Venter, E.; Lennox, C.L.; Meitz-Hopkins, J.C. First report of Cytospora punicae causing post-harvest fruit rot on pomegranate in South Africa. *Plant Dis.* **2017**, *101*, 631. [CrossRef]
19. Munhuweyi, K.; Lennox, C.L.; Meitz-Hopkins, J.C.; Caleb, O.J.; Opara, U.L. Major diseases of pomegranate (*Punica granatum* L.), their causes and management—A review. *Sci. Hortic.* **2016**, *211*, 126–139. [CrossRef]
20. Hussein, Z.; Fawole, O.A.; Opara, U.L. Harvest and postharvest factors affecting bruise damage of fresh fruits. *Hortic. Plant J.* **2019**, *6*, 1–13. [CrossRef]
21. Lufu, R.; Ambaw, A.; Opara, U.L. The contribution of transpiration and respiration processes in the mass loss of pomegranate fruit (cv. Wonderful). *Postharvest Biol. Technol.* **2019**, *157*, 1–10. [CrossRef]
22. Lufu, R.; Ambaw, A.; Opara, U.L. Water loss of fresh fruit: Influencing pre-harvest, harvest and postharvest factors. *Sci. Hortic.* **2020**, *272*, 1–16. [CrossRef]
23. Fadiji, T.; Berry, T.M.; Coetzee, C.J.; Opara, U.L. Mechanical design and performance testing of corrugated paperboard packaging for the postharvest handling of horticultural produce. *Biosyst. Eng.* **2018**, *171*, 220–244. [CrossRef]
24. POMASA (Pomegranate Association of South Africa). Pomegranate Production Guidelines. Available online: https://www.sapomegranate.co.za/wp-content/uploads/2017/08/POMASA-Technical-Production-Manual-1.pdf. (accessed on 29 March 2021).
25. Johnson, L.K.; Dunning, R.D.; Bloom, J.D.; Gunter, C.C.; Boyette, M.D.; Creamer, N.G. Estimating on-farm food loss at the field level: A methodology and applied case study on a North Carolina farm. *Resour. Conserv. Recycl.* **2018**, *137*, 243–250. [CrossRef]
26. Johnson, L.K.; Dunning, R.D.; Gunter, C.C.; Bloom, J.D.; Boyette, M.D.; Creamer, N.G. Field measurement in vegetable crops indicates need for reevaluation of on-farm food loss estimates in North America. *Agric. Syst.* **2018**, *167*, 136–142. [CrossRef]
27. Schneider, F.; Part, F.; Göbel, C.; Nina Langen, N.; Gerhards, C.; Kraus, G.F.; Ritter, G. A methodological approach for the on-site quantification of food losses in primary production: Austrian and German case studies using the example of potato harvest. *Waste Manag.* **2019**, *86*, 106–113. [CrossRef]

28. Fernandez-Zamudio, M.-A.; Barco, H.; Schneider, F. Direct Measurement of Mass and Economic Harvest and Post-Harvest Losses in Spanish Persimmon Primary Production. *Agriculture* **2020**, *10*, 581. [CrossRef]
29. Usanmaz, S.; Kahramanoğlu, I.; Yılmaz, N. Yield and pomological characteristics of three pomegranate (*Punica granatum* L.) cultivars: Wonderful, Acco and Herskovitz. *Am. J. Agric. For.* **2014**, 2, 61–65. [CrossRef]
30. Bryman, A.; Bell, E.; Hirschsohn, P.; Dos Santos, A.; Du Toit, J.; Masenge, A.; Aardt, I.; Wagner, C. *Research Methodology: Business and Management Contexts*; Oxford University Press Southern Africa: Cape Town, South Africa, 2014.
31. Dome, M.M.; Prusty, S. Determination of vegetable postharvest loss in the last-mile supply chain in Tanzania: A lean perspective. *Int. J. Logist. Syst. Manag.* **2017**, *27*, 133–150. [CrossRef]
32. Munhuweyi, K.; Opara, U.L.; Sigge, G. Postharvest losses of cabbages from retail to consumer and the socio-economic and environmental impacts. *Br. Food J.* **2016**, *118*, 286–300. [CrossRef]
33. Opara, U.L.; Al-Ani, M.R. Antioxidant contents of pre-packed fresh-cut versus whole fruit and vegetables. *Br. Food J.* **2010**, *112*, 797–810. [CrossRef]
34. González, A.D.; Frostell, B.; Carlsson-Kanyama, A. Protein efficiency per unit energy and per unit greenhouse gas emissions: Potential contribution of diet choices to climate change mitigation. *Food Policy* **2011**, *36*, 562–570. [CrossRef]
35. Mekonnen, M.M.; Hoekstra, A.Y. The green, blue and grey water footprint of crops and derived crop products. *Hydrol. Earth Syst. Sci.* **2011**, *15*, 1577–1600. [CrossRef]
36. Spiker, M.L.; Hiza, H.A.; Siddiqi, S.M.; Neff, R.A. Wasted food, wasted nutrients: Nutrient loss from wasted food in the United States and comparison to gaps in dietary intake. *J. Acad. Nutr. Diet.* **2017**, *117*, 1031–1040. [CrossRef]
37. Paul, D.K.; Shaha, R.K. Nutrients, vitamins and minerals content in common citrus fruits in the northern region of Bangladesh. *Pak. J. Biol. Sci.* **2004**, *7*, 238–242. [CrossRef]
38. Hanson, C.; Lipinski, B.; Robertson, K.; Dias, D.; Gavilan, I.; Gréverath, P.; Ritter, S.; Fonseca, J.; Van Otterdijk, R.; et al.; World Resource Institute. Food Loss and Waste Accounting and Reporting Standard. 2016. Available online: https://www.wri.org/publication/food-loss-and-waste-accounting-and-reporting-standard (accessed on 29 March 2020).
39. Atanda, S.A.; Pessu, P.O.; Agoda, S.; Isong, I.U.; Ikotun, I. The concepts and problems of post–harvest food losses in perishable crops. *Afr. J. Food Sci.* **2011**, *5*, 603–613.
40. Pomegranate Producers Association of South Africa (POMASA). Pomegranate Industry Overview: Statistics and Information. 2017. Available online: https://www.sapomegranate.co.za/statistics-and-information/pomegranate-industry-overview/ (accessed on 12 March 2020).
41. Bond, R. Carrot Loss during Primary Production: Field Waste and Pack House Waste. Master's Thesis, Hedmark University, Hedmark, Norway, 4 May 2016.
42. Bantayehu, M.; Alemayehu, M.; Abera, M.; Bizuayehu, S. Postharvest Losses Assessment of Tropical Fruits in the Market Chain of North Western Ethiopia. *Food Sci. Qual. Manag.* **2017**, *66*, 1–12.
43. Bhattarai, D.R.; Subedi, G.D.; Gautam, I.P.; Chauhan, S. Posthavest Supply Chain Study of Carrot in Nepal. *Int. J. Hortic.* **2017**, *7*, 239–245. [CrossRef]
44. Weerakkody, P.; Jobling, J.; Infante, M.M.V.; Rogers, G. The effect of maturity, sunburn and the application of sunscreens on the internal and external qualities of pomegranate fruit grown in Australia. *Sci. Hortic.* **2010**, *124*, 57–61. [CrossRef]
45. Yazici, K.; Kaynak, L. Effects of air temperature, relative humidity and solar radiation on fruit surface temperatures and sunburn damage in pomegranate (Punica granatum L. cv. Hicaznar). *Acta Hortic.* **2006**, *818*, 181–186. [CrossRef]
46. Ghanbarpour, E.; Rezaei, M.; Lawson, S. Reduction of cracking in pomegranate fruit after foliar application of humic acid, calcium-boron and kaolin during water stress. *Erwerbs-Obstbau* **2019**, *61*, 29–37. [CrossRef]
47. Opara, L.U.; Studman, C.J.; Banks, N.H. Fruit skin splitting and cracking. *Hortic. Rev.* **1997**, *19*, 217–262.
48. Abd El-Rhman, I.E. Physiological studies on cracking phenomena of pomegranates. *J. Appl. Sci. Res.* **2010**, *6*, 696–703.
49. Singh, A.; Shukla, A.K.; Meghwal, P.R. Fruit Cracking in Pomegranate: Extent, Cause, and Management—A Review. *Int. J. Fruit Sci.* **2020**, *1784074*, S1234–S1253. [CrossRef]
50. Hussein, Z.; Fawole, O.A.; Opara, U.O. Effects of bruising and storage duration on physiological response and quality attributes of pomegranate fruit. *Sci. Hortic.* **2020**, *267*, 1–7. [CrossRef]
51. Hussein, Z.; Fawole, O.A.; Opara, U.L. Preharvest factors influencing bruise damage of fresh fruits–a review. *Sci. Hortic.* **2018**, *229*, 45–58. [CrossRef]
52. Czieczor, L.; Bentkamp, C.; Damerow, L.; Blanke, M. Non-invasive determination of the quality of pomegranate fruit. *Postharvest Biol. Technol.* **2018**, *136*, 74–79. [CrossRef]
53. Swilling, M.; Annecke, E. *Just Transitions: Explorations of Sustainability in an Unfair World*; UCT Press: Cape Town, South Africa, 2012.
54. U.S. DOE (Department of Energy, Energy Information Administration). Method for Calculating Carbon Sequestration by Trees in Urban and Suburban Settings. Voluntary Reporting of Greenhouse Gases, U.S. 1998. Available online: https://www3.epa.gov/climatechange/Downloads/method-calculating-carbon-sequestration-trees-urban-and-suburban-settings.pdf. (accessed on 17 July 2020).
55. Gleick, P.H.; Iwra, M. Basic water requirements for human activities: Meeting basic needs. *Water Int.* **1996**, *21*, 83–92. [CrossRef]
56. Muller, J.C. Pomegranate waste record. In *Personal Communication*; Sonlia Fruit Packhouse: Wellington, South Africa, 2020.

57. Keatinge, J.D.; Waliyar, F.; Jamnadas, R.H.; Moustafa, A.; Andrade, M.; Drechsel, P.; Hughes, J.D.A.; Kadirvel, P.; Luther, K. Relearning old lessons for the future of food—by bread alone no longer: Diversifying diets with fruit and vegetables. *Crop. Sci.* **2010**, *50*, 51–62. [CrossRef]
58. Beretta, C.; Stoessel, F.; Baier, U.; Hellweg, S. Quantifying food losses and the potential for reduction in Switzerland. *Waste Manag.* **2013**, *33*, 764–773. [CrossRef]
59. Medici, M.; Canavari, M.; Toselli, M. Interpreting environmental impacts resulting from fruit cultivation in a business innovation perspective. *Sustainability* **2020**, *12*, 1–14.
60. Wright, H.J.; Ochilo, W.; Pearson, A.; Finegold, C.; Oronje, M.; Wanjohi, J.; Kamau, R.; Holmes, T.; Rumsey, A. Using ICT to strengthen agricultural extension systems for Plant health. *J. Agric. Food Inf.* **2016**, *17*, 23–36. [CrossRef]
61. Gustavsson, J.; Cederberg, C.; Sonesson, U.; Van Otterdijk, R.; Meybeck, A. *Global Food Losses and Food Waste*; FAO: Rome, Italy, 2011; pp. 1–38.
62. Porat, R.; Lichter, A.; Terry, L.A.; Harker, R.; Buzby, J. Postharvest losses of fruit and vegetables during retail and in consumers' homes: Quantifications, causes, and means of prevention. *Postharvest Biol. Technol.* **2018**, *139*, 135–149. [CrossRef]

sustainability MDPI

Article

Postharvest Losses in Quantity and Quality of Table Grape (cv. Crimson Seedless) along the Supply Chain and Associated Economic, Environmental and Resource Impacts

Anelle Blanckenberg [1,2], Umezuruike Linus Opara [2,*] and Olaniyi Amos Fawole [2,3,*]

1 Department of Horticultural Sciences, Faculty of AgriSciences, Stellenbosch University, Stellenbosch 7600, South Africa; 12938033@sun.ac.za
2 Africa Institute of Postharvest Technology, South African Research Chair in Postharvest Technology, Department of Horticultural Sciences, Faculty of AgriSciences, Stellenbosch University, Stellenbosch 7600, South Africa
3 Postharvest Research Laboratory, Department of Botany and Plant Biotechnology, University of Johannesburg, P.O. Box 524, Auckland Park, Johannesburg 2006, South Africa
* Correspondence: opara@sun.ac.za (U.L.O.); olaniyi@sun.ac.za (O.A.F.)

Abstract: High incidence of postharvest losses is a major challenge to global food security. Addressing postharvest losses is a better strategy to increase business efficiency and improve food security rather than simply investing more resources to increase production. Global estimates show that fruit and vegetables are the highest contributors to postharvest losses and food waste, with 45% of production lost. This represents 38% of total global food losses and waste. However, the lack of primary data on postharvest losses at critical steps in the fruit value chain and the unknown economic, environmental and resource impacts of these losses makes it difficult to formulate mitigation strategies. This paper quantifies postharvest losses and quality attributes of 'Crimson Seedless' table grapes at farm and simulated retail levels. Table grapes were sampled from four farms in the Western Cape Province of South Africa, the largest deciduous fruit production and export region in Southern Africa. Mean on-farm losses immediately after harvest was 13.9% in 2017 and 5.97% in 2018, ranging from 5.51% to 23.3% for individual farms. The main reason for on-farm losses was mechanical damage (7.1%). After 14 days in cold storage (-0.3 ± 0.7 °C, 81.3 ± 4.1% RH), mean grape losses were 3.05% in 2017 and 2.41% in 2018, which increased to 7.41% in 2017 and 2.99% in 2018, after 28 days. After 10 days of further storage under simulated market conditions (5.4 ± 0.6 °C, 83.7 ± 2.9% RH), fruit losses were 3.65% during retail marketing and 4.36% during export. Storing grapes under ambient conditions (25.1 ± 1.3 °C and 46.6 ± 6.0% RH) resulted in a higher incidence of losses, increasing from 7.03 to 9.59 and 14.29% after 3, 7 and 10 days, respectively. The socioeconomic impacts of these postharvest losses amounted to financial losses of over ZAR 279 million (USD 17 million according to the conversion rate of 20 October 2020) annually, and this was associated with the loss of 177.43 million MJ of fossil energy, 4.8 million m^3 of fresh water and contributed to the emission of approximately 52,263 tons of CO_2 equivalent.

Keywords: postharvest losses; food waste; physicochemical properties; table grape; shelf-life; decay; stem browning; SO_2 damage; socioeconomic impacts

Citation: Blanckenberg, A.; Opara, U.L.; Fawole, O.A. Postharvest Losses in Quantity and Quality of Table Grape (cv. Crimson Seedless) along the Supply Chain and Associated Economic, Environmental and Resource Impacts. *Sustainability* **2021**, *13*, 4450. https://doi.org/10.3390/su13084450

Academic Editors: Alessandro Suardi and Nadia Palmieri

Received: 8 March 2021
Accepted: 7 April 2021
Published: 16 April 2021

1. Introduction

Interest in the mitigation of postharvest losses is heightened due to global concern about food insecurity. Preserving the food supply after production has since the earliest times been a problem for humankind [1]. However, a dominant challenge of the 21st century is how to feed the growing world population sustainably, predicted to reach 9.1 billion by 2050, affordably while using the natural resources required equitably and sustainably [2,3].

The UN Food and Agriculture Organization (FAO) estimated that feeding the growing world population by 2050 would require 70% increase in food production [4]. However, the FAO also reports that approximately one-third of the edible portions of the food produced globally is lost or wasted along the supply chain, from farm to plate [5], with a total of 38% of the volume consisting of fruits and vegetables [6]. Lost and wasted food consumes a quarter of all the water used by agriculture annually, requires farmland area the size of China, and generates an estimated 8% of global greenhouse gas emissions. In effect, if lost and wasted food were a country, it would be the third-largest greenhouse gas emitter on the planet after China and the United States [7–9].

While 95% of all agricultural research investment focuses on increasing crop production strategies, only 5% focuses on postharvest issues [10]. Reducing postharvest loss and waste is more cost effective and less time consuming than production strategies. Therefore, the improvement of postharvest chains must receive as much attention as production practices. Furthermore, limiting the loss of fresh fruit will reduce the use of land, chemicals, energy, and other inputs needed to produce horticultural crops, thereby conserving natural resources and protecting the environment [11,12].

A major obstacle in achieving postharvest loss mitigation is a lack of clear knowledge of the real magnitudes of losses, making it impossible to measure progress against any loss reduction targets [13].

Apart from worrying anecdotal information, there is little scientific data available about food losses in South Africa. A study [14] provided a preliminary estimate of the magnitude of food loss and waste generation in South Africa of around 9.04 million tons per year. However, this study was based on available data as reported by FAO and further assumptions as no primary data was collected. While estimates do indicate the size of the problem, they do not provide accurate and reliable data for specific supply chains.

The lack of primary data on postharvest losses at critical steps in the fruit value chain and the unknown economic, environmental and resource impacts makes it challenging to formulate mitigation strategies. There is a lack of standard methods to measure postharvest losses of food crops, including fruit and vegetable crops. Researchers have developed many different methods during the past few decades, focusing on different aspects of the value chain and varying types of food losses [15]. The lack of accurate and reliable postharvest loss data may result in inaccurate assumptions on food wastage [16].

In South Africa, fruit and vegetables account for 47% of food wastage [14]. In the first South African study generating primary data on postharvest losses of vegetables at the retail level, the authors [12] found vegetable losses for carrot, tomato and cabbages to be on average 17.93%, 15.33% and 21.21%, respectively. There are currently no primary research data on the magnitude of losses in the fresh fruit value chain.

Fruit is a major contributor to the agricultural industry, considering foreign exchange earnings and employment creation.

Table grapes account for 32% of the total area planted to deciduous fruits in South Africa [17], with 'Crimson Seedless' cultivar accounting for 20% of all vines [18]. 'Crimson Seedless' is a mid to late season red seedless grape cultivar with firm berries characterized by its crisp texture and intensely sweet flavor. The main quality problems during postharvest handling are moisture loss leading to loss of mass, fungal infection and shriveled rachis, which then become brittle and break easily, and the browning of the stems, which reduces the visual appeal and price of the product [19–21].

Although table grapes are nonclimacteric fruit with a relatively low rate of physiological activity [22], they are subject to severe postharvest losses during storage and long distance transport [23]. Rapid moisture loss, which results in rachis (cluster stem) drying and browning [24,25], mass loss [26], berry shatter, wilting and shriveling of berries [27] are some of the main quality problems experienced during postharvest handling causing quantitative and quality losses. It has been suggested that inappropriate handling processes are the main reason for weakening the natural defenses of grapes and making fresh grapes more susceptible to decay and subsequent deterioration [28].

As the gross value of production of table grapes has increased significantly from R2 billion in 2006/07 to R7.1 billion in 2015/16, an increase of 248% [17], reduction of postharvest losses by even a few percentage points will not only reduce the cost of production, trade and distribution but will also have significant financial implications for all involved in the supply chain by lowering the price for the consumer and increasing the farmers' income. Therefore, this research aimed to fill a gap in the knowledge by generating primary data on postharvest losses at critical steps in the supply chain for table grape 'Crimson Seedless' to inform future action to reduce and better manage the food waste problem.

2. Materials and Methods

2.1. Harvesting Techniques and Berry Preparation

Data collection protocols were similar to those used by [29]. Grapes were harvested from four farms during the last week of February and the first week of March in 2017, and during the first three weeks of February in 2018. Both times were during the commercial harvest. In 2017 a total sample size of 1200 bunches (300 bunches per farm) were collected, while 1600 bunches (400 bunches per farm) were collected in 2018. The grapes were collected from four farms that each have their own packhouse on site, near De Doorns, Robertson and Piketberg in the Western Cape Region of South Africa. Bunches were weighed in the packhouse as they came in from the vineyard, each bunch was then tagged with a unique label identifying the farm of origin (V; K; R or D), the supply chain scenario the bunch was destined for (A; B; or C in 2017 and A; B; C or D in 2018) and the bunch number (1–100) e.g., VA29 (Table 1).

Table 1. Description of the supply chain scenarios studied.

Supply Chain Scenario	Description	Environmental Condition
A	Table grapes were harvested and stored under ambient conditions, typical in areas that lack cold storage facilities Measurements were taken at harvest and after 3, 7 and 10 days	Under ambient conditions for 10 days: 25.1 ± 1.3 °C 46.6 ± 6.0% RH
B	Handling of table grapes for domestic supply chain Measurements were taken at harvest, after 14 days in cold storage, after 10 days at retail conditions and then after 3, 7 and 10 days at ambient conditions	Cold store for 2 weeks: −0.3 °C ± 0.7 °C and 81.3% ± 4.1% RH Retail store for 10 days: 5.4 °C ± 0.6 °C and 83.7% ± 2.9% RH Consumer/home (ambient) store: 25.1 ± 1.3 °C and 46.6 ± 6.0%RH
C	Shipping to export markets Measurements were taken at harvest, after 28 days in cold storage, after a further 10 days at retail conditions and then at 3, 7 and 10 days at ambient conditions	Cold storage for 4 weeks at −0.3 ± 0.7 °C, 81.3 ± 4.1% RH Retail store for 10 days: 5.4 °C ± 0.6 °C and 83.7% ± 2.9% RH Consumer/home (ambient) 'shelf' store: 25.1 ± 1.3 °C and 46.6 ± 6.0%RH
D	Reefer container containing export fruit are left open on arrival for 2 days before fruit is unloaded. 'Abusive' treatment of fruit within the export chain Measurements were taken at harvest, after 28 days in cold storage then after 2 days exposure to ambient conditions, after a further 10 days at retail conditions and then at 3, 7 and 10 days at ambient conditions	Cold store for 2 weeks: −0.3 °C ± 0.7 °C and 81.3% ± 4.1% RH; Ambient storage for 2 days: 25.1 ± 1.3 °C, 46.6 ± 6.0% RH; Retail store display for 10 days: 5.4 °C ± 0.6 °C and 83.7% ± 2.9% RH; Consumer/home (ambient) 'shelf' store: 25.1 ± 1.3 °C and 46.6 ± 6.0% RH

The grape bunches were then trimmed by expert packers according to commercial practice and packed into standard 9-kg cartons (internal dimensions of the 9-kg boxes were 58 cm long × 34 cm wide × 13.2 cm high) with a riffled sheet at the bottom, a plastic liner as well as an SO_2 cover pad on top with a slow release of sodium metabisulfite ($Na_2S_2O_5$) (Uvasys®, Cape Town, South Africa). Sodium metabisulfite generates sulphurous anhydride gas ($S0_2$) when in contact with humidity, inhibiting the development and growth of fungi in table grapes during refrigerated packaging and transport.

2.2. Supply Chains Simulated

In 2017, 18 cartons per farm were collected and divided equally into three simulated supply chain scenarios. In 2018, 24 cartons per farm were collected and divided into four simulated supply chain scenarios. From each farm and for both years, 100 bunches were used for each supply chain scenario simulated, i.e., 400 bunches per scenario. Four supply chain scenarios were studied (Table 1), representing the range of postharvest handling practices that occur in local and export marketing of table grapes in the South African fresh fruit industry. According to export grape producers (pers. communication Amelia Vorster, Technical Advisor (Quality)—Karsten Western Cape), scenario D is a common occurrence and leads to tension between role-players as to whether the fruit was mishandled before the report was written and who is responsible for the losses if it is higher than expected.

2.3. Fruit Loss Evaluation and Quality Measurements

2.3.1. Postharvest Losses

The base measurement for losses at harvest occurred in the packhouse on each farm after the bunches were trimmed for packaging and the resulting berries sorted into categories based on the reason for being cut from the bunch, (1) berry too green in color, (2) mechanical damage or (3) decayed. It was quantified as the weight of the berries removed as a percentage of the original weight of the bunches before trimming. At each evaluation date thereafter, physical losses were quantified as the decrease in bunch weight and the amount lost due to decay or SO_2 damage (a high concentration of SO_2 can damage table grapes by causing bleaching, cracking or causing early browning of the rachis) expressed as a percentage of the initial berry numbers per bunch.

2.3.2. Quality Attributes

The following attributes were measured at each evaluation time:

1. Weight loss

Expressed as a percentage of the initial bunch weight. 30 bunches × 4 farms = 120 bunches per supply chain scenario.

2. Stem browning

Rated on a 5-point scale, with 1 being fresh/green and 5 being dry/brown [15,16]. In total, 30 bunches × 4 farms = 120 bunches per supply chain scenario.

3. Total soluble solids (TSS) concentration

Fruit juice was extracted using a manual juice extractor (TMS® hand press commercial pro manual juice squeezer). TSS of juice was measured with a digital refractometer (Atago, Tokyo, Japan). A total of 12 bunches per supply chain scenario were used.

4. Titratable acidity (TA)

TA of juice was determined by titration to pH 8.2 using a Metrohm 862 compact titrosampler (Herisau, Switzerland). A total of 12 bunches per supply chain scenario were used.

5. Peel color

Color was assessed using a colorimeter (Minolta CR-400, Minolta Corp, Osaka, Japan) and expressed as CIE L*, a*, b* coordinate where L* defines lightness, a* denotes the red/green value and b* the yellow/blue value [30]. A total of, 120 berries per supply chain scenario were evaluated for peel color.

6. Firmness

Berry firmness (N) was measured by compression (TA.XT.plus, Stable Micro Systems Ltd., Surrey, UK) [31]. In total, 120 berries per supply chain scenario were evaluated for firmness.

2.4. Environmental and Economic Impacts of Postharvest Losses

Total greenhouse gas emissions were calculated using values provided by [32]. That study examined the annual cycle for grape production, beginning with establishment costs, raw material extraction for production of inputs used on the vineyard and included the factors of fertilizer, tillage, irrigation, pest management, electricity and fuel consumption, ending at delivery of grapes. For every ton of grapes produced, stored and transported to the retail market, approximately 0.91 ton of CO_{2eq} is emitted into the atmosphere. The energy cost for producing and marketing the lost produce was obtained using a reference value of 6529 MJ/ton provided by [33], and the water footprint was determined by multiplying the quantity of lost produce with the reference water footprint value of 210.35 m^3/ton provided by [34]. The value of table grapes lost was calculated using values provided by [17], R13134/ton for locally sold produce and R21002/ton for exported produce.

2.5. Statistical Analysis

Data on farm losses at harvest were subjected to a one-way analysis of variance (ANOVA) and the physicochemical analysis data for firmness, total soluble solids (TSS), titratable acidity (TA), peel color, weight loss, decay, SO_2 damage, and stem browning were subjected to mixed model analysis of variance (ANOVA) using Statistica version 13.2 (TIBCO Software Inc., Palo Alto, CA, USA) with 'farm' and 'time' as fixed effect and cartons as a random effect.

3. Results

3.1. Physical Losses at Farm Level

In 2017 the measured losses at harvest for individual farms were 7.5%, 9.7%, 15.7%, and 23.3% for V, K, R, or D, respectively, while in 2018 the same farms lost 6.17%, 6.39%, 5.51%, and 5.85%. The average loss in 2017 was 13.9% and 5.97% in 2018. The main reasons for the losses in 2017 were mechanical damage (7.1%), poor berry color (5%), and decay (1.8%). In 2018 the reasons remained the same, although the amounts lost differed with mechanical damage (3.09%), poor berry color (1.77%), and decay (1.11%).

3.2. Physical Losses along the Simulated Supply Chain

3.2.1. Weight Loss, Decay and SO_2 Damage

Supply chain scenario A (handling and marketing fruit under ambient conditions)

There was no statistically significant weight loss (p = 0.28) after harvest in 2017 (Table 2); however, there was a decrease in weight of 2.34% after 3 days, 4.47% after 7 days and 7.6% after 10 days, while in 2018 a decrease of 1.55% was noted after 3 days, 1.83% after 7 days and 4.43% after 10 days. (p = 0.19). While not statistically significant, this decrease in weight is important in terms of losses as it could affect the profit margin as fruit are sold by weight. The incidence of decay increased significantly ($p < 0.01$) over time from 1% after 3 days to 3.3% after 7 days and 7.6% after 10 days in 2017 and from 0.85% after 3 days to 1.67% after 7 days and 2.67% 10 days after harvest in 2018 ($p < 0.01$).

Table 2. Physical losses of 'Crimson Seedless' table grapes measured as weight loss (%) and decayed and SO_2 damaged berries (%) after 3, 7 and 10 days at ambient conditions (25.1 ± 1.3 °C and 46.6 ± 6.0%RH). Mean values with different letter(s) in the same column indicate statistically significant differences ($p < 0.05$).

Season	2017			2018		
Time	Weight Loss (%)	Decay (%)	SO_2 (%)	Weight Loss (%)	Decay (%)	SO_2 (%)
Harvest	-	0	0	-	0	0
3 days	2.34 ^{a}	1.05 ^{b}	1.85 ^{b}	1.55 ^{a}	0.85 ^{b}	0.37 ^{b}
7 days	4.47 ^{a}	3.34 ^{b}	2.31 ^{a}	1.83 ^{a}	1.67 ^{b}	0.92 ^{a}
10 days	7.63 ^{a}	7.60 ^{a}	2.57 ^{a}	4.43 ^{a}	2.67 ^{a}	1.19 ^{a}
p-value	0.28	<0.01	<0.01	0.19	<0.01	<0.01

Note: Mean values within the same column with different letters are significantly ($p < 0.05$) different by Duncan's Multiple Range test (DMRT).

The incidence of SO_2 berry damage increased significantly over time from 1.85% after 3 days to 2.31% after 7 days and 2.57% after 10 days in 2017 and from 0.37% after 3 days to 0.92% after 7 days and 1.19% after 10 days in 2018 ($p < 0.01$).

Supply chain scenario B (to local retail markets)

There was no statistically significant weight loss (Table 3) in both seasons (2017; $p = 0.91$, 2018; $p = 0.99$). However, the weight decreased by 1.41% after 14 days in cold storage, 1.87% after 10 days at retail conditions, and then by 2.53%, 3.78 and 5.36% after 3, 7 and 10 days under ambient conditions, respectively, in 2017. While in 2018, the decrease in weight was 1.85% after 14 days in cold storage, 2.57% after 10 days at retail conditions, and then 4.03%, 4.40 and 6.76% after 3, 7 and 10 days under ambient conditions. The incidence of berry decay increased significantly ($p < 0.01$) over time, although it remained at zero for the 14 days duration in cold storage (−0.3 ± 0.7 °C, 81.3 ± 4.1% RH) in 2017 with a small incidence of 0.4% in 2018. After 10 days at local retail conditions (5.4 ± 0.6 °C, 83.7 ± 2.9% RH), there was a decay incidence of 2.1% in 2017 and 1.2% in 2018. After 3 days under ambient conditions (25.1 ± 1.3 °C and 46.6 ± 6.0%RH), the incidence of decay increased significantly ($p < 0.01$) to 2.5% in 2017 and 2.2% in 2018. After 7 days, this increased to 5.5% in 2017 and 3.3% in 2018, and after 10 days to 8.6% and 4.7% in 2017 and 2018, respectively.

Table 3. Physical losses of 'Crimson Seedless' table grapes measured as weight loss (%) and decayed and SO_2 damaged berries (%) after 14 days cold storage (−0.3 ± 0.7 °C, 81.3 ± 4.1% RH), after another 10 days at retail conditions (5.4 ± 0.6 °C, 83.7 ± 2.9% RH) and then 3, 7 and 10 days at ambient conditions (25.1 ± 1.3 °C and 46.6 ± 6.0%RH). Mean values with different letter(s) in the same column indicate statistically significant differences ($p < 0.05$).

Season	2017			2018		
Time	Weight Loss (%)	Decay (%)	SO_2 (%)	Weight Loss (%)	Decay (%)	SO_2 (%)
Harvest	-	0	0	-	0	0
14 days (−0.5 °C)	1.41 ^{a}	0	1.0 ^{b}	1.85 ^{a}	0.4 ^{e}	0.8 ^{b}
10 days (5 °C)	1.87 ^{a}	2.1 ^{c}	1.8 ^{ab}	2.57 ^{a}	1.2 ^{d}	1.7 ^{ab}
3 days	2.53 ^{a}	2.5 ^{c}	2.1 ^{a}	4.03 ^{a}	2.2 ^{c}	2.1 ^{a}
7 days	3.78 ^{a}	5.5 ^{b}	2.2 ^{a}	4.40 ^{a}	3.3 ^{b}	2.1 ^{a}
10 days	5.36 ^{a}	8.6 ^{a}	2.4 ^{a}	6.76 ^{a}	4.7 ^{a}	2.1 ^{a}
p-value	0.91	<0.01	0.02	0.99	<0.01	<0.01

Note: Mean values within the same column with different letters are significantly ($p < 0.05$) different by Duncan's Multiple Range test (DMRT).

The incidence of SO_2 berry damage in this study was 1% after 14 days in cold storage in 2017 ($p = 0.02$) and 0.8% in 2018 ($p < 0.01$). SO_2 damage increased significantly after

10 days at local retail conditions from 1.8% in 2017 and 1.7% in 2018 to 2.1% after 3 days at ambient conditions in both years, after which there was no further significant change.

Supply chain scenario C (to international retail markets)

In 2017, weight decreased by 4.82% after 28 days in cold storage, 5.50% after 10 days at retail conditions and 6.61%, 7.90% and 10.22% after 3, 7 and 10 days under ambient conditions, respectively. In 2018, percentage decreases in weight were 1.89% after 28 days in cold storage, 2.45% after 10 days at retail conditions and 2.64%, 3.95% and 5.18% after 3, 7 and 10 days under ambient conditions, respectively. (Table 4). Percentage decay increased significantly over time ($p < 0.01$). After 28 days in cold storage (−0.3 ± 0.7 °C, 81.3 ± 4.1% RH), there was a 2.14% incidence of decay in 2017 and 0.94% in 2018. After 10 days at retail display conditions (5.4 ± 0.6 °C, 83.7 ± 2.9% RH) this increased to 3.2% in 2017 and 2.6% in 2018. After being moved to ambient temperature and humidity conditions (25.1 ± 1.3 °C and 46.6 ± 6.0%RH) for 3 days, decay increased to 4.44% in 2017 and 3.16% in 2018. After 7 days, decay increased to 6.53% in 2017 and 4.95% in 2018, and after 10 days, it reached 9.92% in 2017 and 8.30% in 2018. SO_2 damage remained low. After 28 days in cold storage plus 10 days at retail conditions, less than 0.5% damage was visible. In 2017 it increased significantly ($p < 0.01$) after removal from cold storage to 1.39% after 3 days at ambient conditions, 1.68% after 7 days and 1.85% after 10 days, but remained below 2% overall. In 2018, however, it did not increase significantly ($p = 0.26$) over time and remained below 1% overall.

Table 4. Physical losses of 'Crimson Seedless' table grapes measured as weight loss (%), decayed and SO_2 damaged berries (%) after 28 days cold storage (−0.3 ± 0.7 °C, 81.3 ± 4.1% RH), after another 10 days at retail conditions (5.4 ± 0.6 °C, 83.7 ± 2.9% RH) and then for 3, 7 and 10 days at ambient conditions (25.1 ± 1.3 °C and 46.6 ± 6.0%RH). Mean values with different letter(s) in the same column indicate statistically significant differences ($p < 0.05$).

Season	2017			2018		
Time	**Weight Loss (%)**	**Decay (%)**	**SO_2 (%)**	**Weight Loss (%)**	**Decay (%)**	**SO_2 (%)**
Harvest	-	0	0	-	0	0
28 days (−0.5 °C)	4.82 ^{a}	2.14 ^{d}	0.47 ^{c}	1.89	0.94 ^{d}	0.25 ^{a}
10 days (5 °C)	5.50 ^{a}	3.20 ^{cd}	0.94 ^{c}	2.45	2.60 ^{c}	0.62 ^{a}
3 days	6.61 ^{a}	4.44 ^{c}	1.39 ^{b}	2.64	3.16 ^{c}	0.62 ^{a}
7 days	7.90 ^{a}	6.53 ^{b}	1.68 ^{ab}	3.95	4.95 ^{b}	0.62 ^{a}
10 days	10.22 ^{a}	9.92 ^{a}	1.85 ^{a}	5.18	8.30 ^{a}	0.62 ^{a}
p-value	0.85	<0.01	<0.01	0.84	<0.01	0.26

Note: Mean values within the same column with different letters are significantly ($p < 0.05$) different by Duncan's Multiple Range test (DMRT).

Supply chain scenario D (simulated 'abusive' treatment of fruit within the export chain)

There was no statistically significant difference in bunch weight over time ($p = 0.19$), although a 1.35% decrease in weight is noted after 28 days in cold storage, 2.17% after 2 days at 'abusive' ambient conditions, 3.04% after 10 days at retail conditions and 3.96%, 4.74% and 5.70% after 3, 7 and 10 days under ambient conditions, respectively (Table 5). Percentage decay increased significantly over time ($p < 0.01$). After 28 days in cold storage, there was a 0.84% incidence of decay, increasing to 1.34% after breaking the cold chain with 2 days at 'abusive' ambient conditions, 10 days at retail conditions increased this to 2.35%, and decay kept increasing significantly at ambient conditions to 3.24% after 3 days, 4.46% after 7 days and 6.7% after 10 days. SO_2 damage remained low, with only 0.88% visible after 28 days in cold storage and did not increase significantly ($p = 0.99$) over time, remaining around 1%.

Table 5. Physical losses of 'Crimson Seedless' table grapes measured as weight loss (%), decayed and SO_2 damaged berries (%) after 28 days cold storage (−0.3 ± 0.7 °C, 81.3 ± 4.1% RH), 2 days 'abusive' temperature and humidity (25.1 ± 1.3 °C and 46.6 ± 6.0%RH), another 10 days at retail conditions (5.4 ± 0.6 °C, 83.7 ± 2.9% RH) and then for 3, 7 and 10 days at ambient conditions (25.1 ± 1.3 °C and 46.6 ± 6.0%RH). Mean values with different letter(s) in the same column indicate statistically significant differences ($p < 0.05$).

Season		2018	
Time	**Weight Loss (%)**	**Decay (%)**	**SO_2 (%)**
Harvest	-	0	0
28 days (−0.5 °C)	1.31 [a]	0.84 [e]	0.88 [a]
2 days (ambient)	2.17 [a]	1.33 [e]	0.98 [a]
10 days (5 °C)	3.04 [a]	2.35 [d]	1.06 [a]
3 days	3.96 [a]	3.24 [c]	1.09 [a]
7 days	4.74 [a]	4.46 [b]	1.10 [a]
10 days	5.70 [a]	6.70 [a]	1.21 [a]
p-value	0.19	<0.01	0.99

Note: Mean values within the same column with different letters are significantly ($p < 0.05$) different by Duncan's Multiple Range test (DMRT).

3.2.2. Total Amount of Physical Losses

When the amount of weight loss, decay and SO_2 damage are combined, the total amount of losses along the different supply chain scenarios are as follows (Figures 1 and 2):

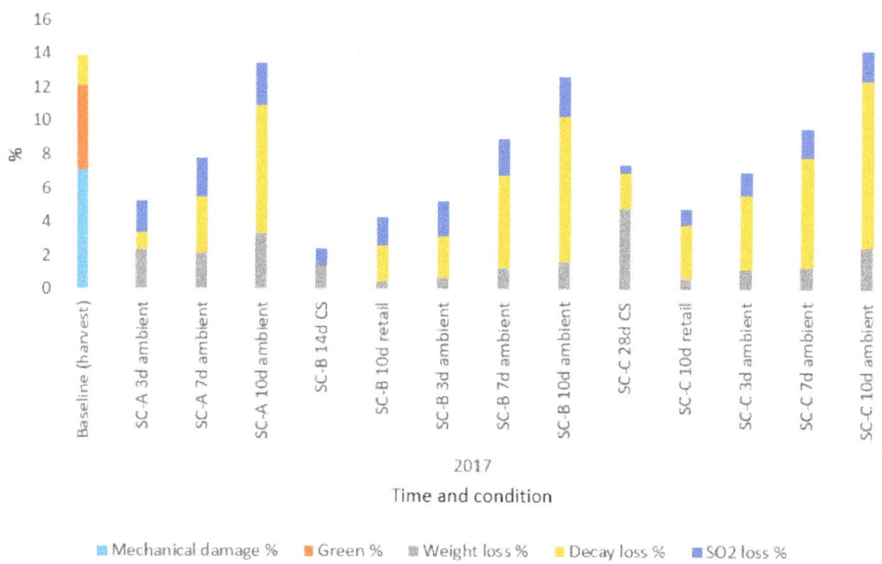

Figure 1. Total physical losses of grape 'Crimson Seedless' at harvest 2017 along different supply chain (SC) scenarios where SC-A represents marketing at ambient conditions, SC-B represents the supply chain to the local retail market, SC-C represents the international supply chain, and SC-D represents the international supply chain including 2 days 'abusive' ambient temperature and humidity.

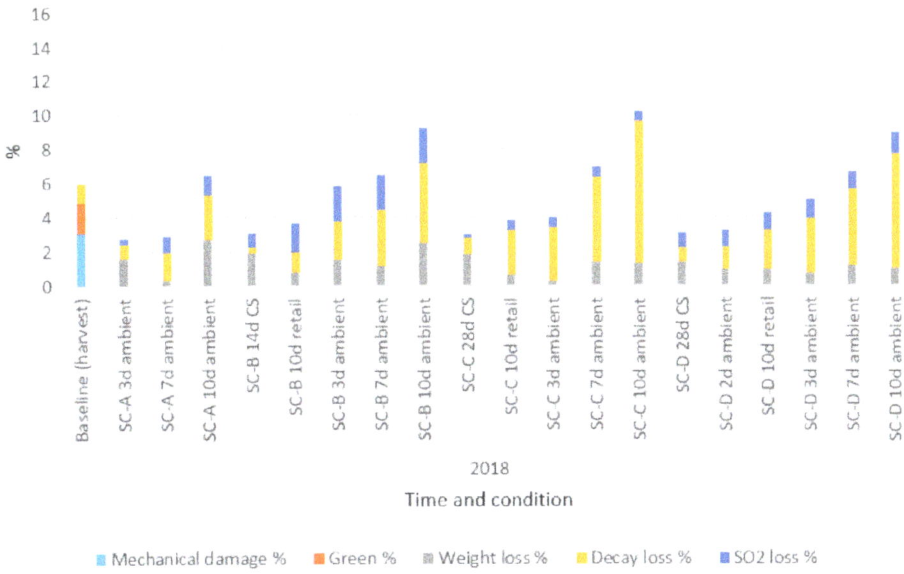

Figure 2. Total physical losses of grape 'Crimson Seedless' at harvest 2018 along different supply chain (SC) scenarios where SC-A represents marketing at ambient conditions, SC-B represents the supply chain to the local retail market, SC-C represents the international supply chain, and SC-D represents the international supply chain including 2 days 'abusive' ambient temperature and humidity.

Supply chain scenario A (marketing at ambient conditions)

In 2017, 13.9% was lost at harvest, followed by 5.25% after 3 days, 7.83% after 7 days, and 13.47% after 10 days. The total losses were 27.37%. In 2018 only 5.97% were lost at harvest, 2.77% after 3 days, 2.87% after 7 days and 6.51% after 10 days. The total losses were 12.48%, less than half that of the previous season.

Supply chain scenario B (to local retail markets)

After 14 days in cold storage in 2017, 2.41% were lost, 4.36% after 10 days at retail condition followed by an additional 5.27% after 3 days at ambient conditions (shelf-life), 8.98% after 7 days, and 12.64% after 10 days. When this is added to the initial 13.9% lost at harvest, the total for this supply chain simulation is 26.54%. In 2018, 3.05% were lost after 14 days in cold storage, 3.65% after 10 days under retail conditions, 5.80% after 3 days under ambient conditions and remained thus after 7 days while increasing again to 10.22% after 10 days. The total loss along this simulated supply chain in 2018 was 16.19%.

Supply chain scenario C (to international markets)

In 2017, 7.41% were lost after 28 days in cold storage, another 4.36% after 10 days under retail conditions and a further 7.03% after 3 days under ambient conditions, 9.59% after 7 days and 14.29% after 10 days. A total of 28.19% was lost in this export supply chain simulation in 2017. In 2018, 2.99% were lost after 28 days in cold storage, 3.79% after 10 days under retail conditions, 3.97% after 3 days under ambient conditions, 6.92% after 7 days, and 10.22% after 10 days. A total of 16.99% were lost in 2018.

Supply chain scenario D (simulated 'abusive' storage conditions of fruit within the international supply chain)

This simulation was done during the 2018 season only. It was found that 3.03% was lost after 28 days in cold storage, 3.19% after 2 days of 'abusive' ambient condition, 4.27% after 10 days at retail conditions. After 3 days under ambient condition losses increased to 5%, 6.66% after 7 days, and 8.91% after 10 days.

3.3. *Quality Losses along the Supply Chain*

Supply chain scenario A (marketing at ambient conditions)

In the 2017 season, the berry color became lighter (L) over time, (Table 6) although the change was just not statistically significant ($p = 0.06$), during the 2018 season (Table 7), however, the change in lightness (L) of berry color became statistically significant ($p < 0.01$). The measurements for a* denoting the red/green values ($p = 0.45$) and b* indicating the yellow/blue values ($p = 0.98$) in 2017 and a* ($p = 0.21$) and b* ($p = 0.75$) in 2018, did not change significantly. There was no significant difference in firmness in 2017 or 2018 ($p = 0.79$), although the values decreased over time. The TSS ($p = 0.85$) and TA ($p = 0.50$) values did not change significantly in 2017 or in 2018, TSS ($p = 0.75$) and TA ($p = 0.73$). For both seasons, the stem color changed from fresh and green to mostly dry and brown within 7 days after harvest ($p < 0.01$).

Table 6. Supply chain scenario A (2017): changes in quality attributes of color (L, a* and b*), firmness (N), total soluble solids (TSS) (Brix °), titratable acidity (TA) (%), and stem browning index for 'Crimson Seedless' table grapes at harvest and after 3, 7 and 10 days at ambient conditions (25.1 ± 1.3 °C and 46.6 ± 6.0%RH). Mean values with different letter(s) in the same column indicate statistically significant differences ($p < 0.05$).

Year						**2017**	
Time	**L**	**a***	**b***	**Firmness (N)**	**TSS (Brix °)**	**TA (%)**	**Stem Browning Index**
Harvest	29.55 ^{a}	5.85 ^{a}	6.77 ^{a}	98.22 ^{a}	18.72 ^{a}	0.89 ^{a}	1 ^{d}
3 days	30.81 ^{a}	4.98 ^{a}	6.56 ^{a}	97.80 ^{a}	18.16 ^{a}	0.75 ^{a}	2.4 ^{c}
7 days	31.07 ^{a}	5.08 ^{a}	6.59 ^{a}	96.87 ^{a}	18.34 ^{a}	0.73 ^{a}	4.4 ^{b}
10 days	30.91 ^{a}	4.88 ^{a}	6.75 ^{a}	95.47 ^{a}	19.15 ^{a}	0.80 ^{a}	5.0 ^{a}
p-value	0.06	0.45	0.98	0.79	0.85	0.50	<0.01

Note: Mean values within the same column with different letters are significantly ($p < 0.05$) different by Duncan's Multiple Range test (DMRT).

Table 7. Supply chain scenario A (2018): changes in quality attributes of color (L, a* and b*), firmness (N), TSS (Brix °), TA (%), and stem browning index for 'Crimson Seedless' table grapes at harvest and after 3, 7 and 10 days at ambient conditions (25.1 ± 1.3 °C and 46.6 ± 6.0%RH). Mean values with different letter(s) in the same column indicate statistically significant differences ($p < 0.05$).

Year						**2018**	
Time	**L**	**a***	**b***	**Firmness (N)**	**TSS (Brix °)**	**TA (%)**	**Stem Browning Index**
Harvest	27.46 ^{c}	7.91 ^{a}	6.95 ^{a}	98.15 ^{a}	20.88 ^{a}	0.86 ^{a}	1 ^{c}
3 days	27.64 ^{bc}	6.80 ^{a}	6.74 ^{a}	97.73 ^{a}	16.77 ^{a}	0.76 ^{a}	2.5 ^{b}
7 days	27.96 ^{b}	6.84 ^{a}	6.83 ^{a}	96.96 ^{a}	18.60 ^{a}	0.79 ^{a}	4.6 ^{a}
10 days	28.94 ^{a}	7.85 ^{a}	7.00 ^{a}	95.55 ^{a}	19.15 ^{a}	0.85 ^{a}	4.8 ^{a}
p-value	0.06	0.45	0.98	0.79	0.85	0.50	<0.01

Note: Mean values within the same column with different letters are significantly ($p < 0.05$) different by Duncan's Multiple Range test (DMRT).

Supply chain scenario B (to local retail markets)

In 2017, there were no significant changes in any color attributes for lightness (L) denoting black/white values ($p = 0.79$), a* denoting the red/green values ($p = 0.49$) or b* indicating the yellow/blue values ($p = 0.25$) (Table 8). In 2018, no significant differences were measured for a* ($p = 0.31$) and b* ($p = 0.19$). However, lightness (L) values increased ($p < 0.01$), although this only became significant after 10 days at ambient conditions as there was no significant difference between baseline measurements, 14 days in cold storage, 10 days at retail conditions, or even after a week at ambient temperature and humidity (Table 9). Berry firmness ($p = 0.21$) in 2017 and ($p = 0.49$) in 2018, showed no statistically significant changes. While TA ($p = 0.27$) and TSS ($p = 0.73$) showed no significant changes in 2017, in 2018 the values did indicate an increase for both TA ($p < 0.01$) and TSS ($p < 0.01$) over the storage period. For both seasons, bunch stems and rachi remained fresh and green during the 14 days in cold storage (−0.3 ± 0.7 °C, 81.3 ± 4.1% RH), but changed significantly ($p < 0.01$) during 10 days at retail display conditions (5.4 ± 0.6 °C, 83.7 ± 2.9% RH), and after 7 days storage under ambient conditions (25.1 ± 1.3 °C and 46.6 ± 6.0% RH), the stems were mostly dry and brown.

Table 8. Supply chain scenario B (2017): changes in quality attributes of color (L, a* and b*), firmness (N), TSS (Brix °), TA (%), and stem browning index for 'Crimson Seedless' table grapes at harvest after 14 days cold storage (−0.3 ± 0.7 °C, 81.3 ± 4.1% RH), after another 10 days at retail conditions (5.4 ± 0.6 °C, 83.7 ± 2.9% RH) and then for 3, 7 and 10 days at ambient conditions (25.1 ± 1.3 °C and 46.6 ± 6.0%RH). Mean values with different letter(s) in the same column indicate statistically significant differences ($p < 0.05$).

				2017			
Time	**L**	**a***	**b***	**Firmness (N)**	**TSS (Brix °)**	**TA (%)**	**Stem Browning Index**
Harvest	33.65 [a]	7.76 [a]	7.28 [a]	98.23 [a]	17.34 [a]	0.99 [a]	1 [d]
14 days (−0.5 °C)	30.78 [a]	6.60 [a]	7.08 [a]	100.14 [a]	19.39 [a]	0.91 [a]	1.4 [d]
10 days (5 °C)	29.78 [a]	7.64 [a]	7.74 [a]	106.59 [a]	18.97 [a]	0.75 [a]	2.4 [c]
3 days	30.61 [a]	6.11 [a]	7.41 [a]	92.02 [a]	19.63 [a]	0.77 [a]	3.5 [b]
7 days	30.52 [a]	6.70 [a]	8.44 [a]	87.59 [a]	19.52 [a]	0.70 [a]	4.7 [a]
10 days	31.38 [a]	6.86 [a]	9.20 [a]	88.88 [a]	17.10 [a]	0.68 [a]	4.9 [a]
p-value	0.79	0.49	0.24	0.21	0.73	0.27	<0.01

Note: Mean values within the same column with different letters are significantly ($p < 0.05$) different by Duncan's Multiple Range test (DMRT).

Table 9. Supply chain scenario B (2018): changes in quality attributes of color (L, a* and b*), firmness (N), TSS (Brix °), TA (%), and stem browning index for 'Crimson Seedless' table grapes at harvest after 14 days cold storage (−0.3 ± 0.7 °C, 81.3 ± 4.1% RH), after another 10 days at retail conditions (5.4 ± 0.6 °C, 83.7 ± 2.9% RH) and then for 3, 7 and 10 days at ambient conditions (25.1 ± 1.3 °C and 46.6 ± 6.0%RH). Mean values with different letter(s) in the same column indicate statistically significant differences ($p < 0.05$).

				2018			
Time	**L**	**a***	**b***	**Firmness (N)**	**TSS (Brix °)**	**TA (%)**	**Stem Browning Index**
Harvest	27.46 [b]	7.91 [a]	2.23 [a]	111.06 [a]	17.84 [b]	0.72 [c]	1 [d]
14 days (−0.5 °C)	27.88 [b]	8.22 [a]	2.87 [a]	114.59 [a]	17.99 [b]	0.85 [b]	1.3 [d]
10 days (5 °C)	27.32 [b]	7.32 [a]	4.05 [a]	121.69 [a]	19.14 [a]	0.82 [b]	2.7 [c]
3 days	27.84 [b]	7.05 [a]	2.19 [a]	115.23 [a]	18.95 [a]	0.78 [c]	3.7 [b]
7 days	27.56 [b]	6.56 [a]	1.94 [a]	119.03 [a]	19.13 [a]	0.89 [a]	4.6 [a]
10 days	30.30 [a]	6.01 [a]	3.45 [a]	120.27 [a]	18.95 [a]	0.85 [b]	4.8 [a]
p-value	<0.01	0.31	0.19	0.49	<0.01	<0.01	<0.01

Note: Mean values within the same column with different letters are significantly ($p < 0.05$) different by Duncan's Multiple Range test (DMRT).

Supply chain scenario C (to international retail markets)

No significant difference in any color attributes (Table 10) was observed in 2017, L ($p = 0.12$), a* ($p = 0.15$) and b* ($p = 0.72$). In 2018, however, the attribute for lightness (L) changed significantly ($p < 0.01$), with berries becoming a bit lighter after removal from cold storage but darkening again after 7 days under ambient conditions (Table 11), while a* ($p = 0.26$) and b* ($p = 0.22$) remained the same. The average berry firmness remained unchanged in 2017 ($p = 0.90$) and in 2018 ($p = 0.11$). No changes were observed in TSS in 2017 ($p = 0.67$) or in 2018 ($p = 0.30$). There is a trend, however, indicating that TA may decrease somewhat over time, but with a p-value of 0.06, it was just not statistically significant in 2017, while it was significant in 2018 ($p < 0.01$). Stem color exhibited the same pattern for both years, remaining mostly fresh and green during the 28 days cold storage (−0.3 ± 0.7 °C, 81.3 ± 4.1% RH) only changing significantly ($p < 0.01$) during the 10 days at retail display conditions (5.4 ± 0.6 °C, 83.7 ± 2.9% RH) to mostly green with some smaller stems that have turned brown. After 3 days at ambient temperature and humidity (25.1 ± 1.3 °C and 46.6 ± 6.0%RH), most of the smaller stems (rachi) are brown, but the main stem is still green, after 7 days, however, most stems are dry and brown.

Table 10. Supply chain scenario C (2017): changes in quality attributes of color (L, a* and b*), firmness (N), TSS (Brix °), TA (%), and stem browning index for 'Crimson Seedless' table grapes at harvest after 28 days cold storage (−0.3 ± 0.7 °C, 81.3 ± 4.1% RH), after another 10 days at retail conditions (5.4 ± 0.6 °C, 83.7 ± 2.9% RH) and then for 3, 7 and 10 days at ambient conditions (25.1 ± 1.3 °C and 46.6 ± 6.0%RH). Mean values with different letter(s) in the same column indicate statistically significant differences ($p < 0.05$).

				2017			
Time	**L***	**a***	**b***	**Firmness (N)**	**TSS (Brix °)**	**TA (%)**	**Stem Browning Index**
Harvest	30.04 ^{a}	10.37 ^{a}	9.36 ^{a}	98.69 ^{a}	17.80 ^{a}	1.13 ^{a}	1 ^{e}
28 days (−0.5 °C)	29.94 ^{a}	8.80 ^{a}	9.16 ^{a}	94.41 ^{a}	18.56 ^{a}	0.74 ^{a}	1.14 ^{e}
10 days (5 °C)	28.61 ^{a}	7.89 ^{a}	9.54 ^{a}	98.70 ^{a}	19.15 ^{a}	0.64 ^{a}	1.97 ^{d}
3 days	31.41 ^{a}	6.81 ^{a}	8.72 ^{a}	97.00 ^{a}	19.28 ^{a}	0.73 ^{a}	3.08 ^{c}
7 days	31.78 ^{a}	6.52 ^{a}	8.10 ^{a}	99.70 ^{a}	19.17 ^{a}	0.67 ^{a}	4.53 ^{b}
10 days	31.69 ^{a}	5.76 ^{a}	8.10 ^{a}	97.20 ^{a}	18.57 ^{a}	0.67 ^{a}	5.00 ^{a}
p-value	0.12	0.15	0.72	0.90	0.67	0.06	<0.01

Note: Mean values within the same column with different letters are significantly ($p < 0.05$) different by Duncan's Multiple Range test (DMRT).

Table 11. Supply chain scenario C (2018): changes in quality attributes of color (L, a* and b*), firmness (N), TSS (Brix °), TA (%), and stem browning index for 'Crimson Seedless' table grapes at harvest after 28 days cold storage (−0.3 ± 0.7 °C, 81.3 ± 4.1% RH), after another 10 days at retail conditions (5.4 ± 0.6 °C, 83.7 ± 2.9% RH) and then for 3, 7 and 10 days at ambient conditions (25.1 ± 1.3 °C and 46.6 ± 6.0%RH). Mean values with different letter(s) in the same column indicate statistically significant differences ($p < 0.05$).

				2018			
Time	**L***	**a***	**b***	**Firmness (N)**	**TSS (Brix °)**	**TA (%)**	**Stem Browning Index**
Harvest	27.45 ^{b}	7.91 ^{a}	2.23 ^{a}	106.60 ^{a}	17.80 ^{a}	0.72 ^{b}	1 ^{e}
28 days (−0.5 °C)	27.75 ^{b}	7.06 ^{a}	1.76 ^{a}	106.02 ^{a}	18.69 ^{a}	0.77 ^{a}	1.97 ^{e}
10 days (5 °C)	28.09 ^{ab}	7.20 ^{a}	2.47 ^{a}	104.34 ^{a}	18.58 ^{a}	0.77 ^{a}	2.67 ^{d}
3 days	28.42 ^{a}	6.96 ^{a}	2.36 ^{a}	110.07 ^{a}	18.77 ^{a}	0.72 ^{b}	3.61 ^{c}
7 days	27.30 ^{c}	6.60 ^{a}	2.10 ^{a}	110.21 ^{a}	18.55 ^{a}	0.69 ^{c}	4.45 ^{b}
10 days	27.79 ^{b}	7.23 ^{a}	2.41 ^{a}	112.65 ^{a}	18.53 ^{a}	0.73 ^{b}	4.78 ^{a}
p-value	<0.01	0.26	0.22	0.11	0.30	$p < 0.01$	<0.01

Note: Mean values within the same column with different letters are significantly ($p < 0.05$) different by Duncan's Multiple Range test (DMRT).

Supply chain scenario D (simulated 'abusive' treatment of fruit within the export chain)

Significant changes ($p < 0.01$) in color attribute (L) for lightness were observed, (Table 12), with berries becoming lighter with increased temperature and lower humidity and darker when returned to lower temperatures and increased humidity. No significant difference in color attributes a* ($p = 0.86$) and b* ($p = 0.21$) were observed. The average berry firmness remained unchanged ($p = 0.23$). There were significant changes observed in TSS ($p < 0.01$) with values increasing and TA values ($p < 0.01$) that decreased during storage. Stem color changed significantly over time ($p < 0.01$). While stems remained fresh and green during the 28 days in cold storage, the 2 days at ambient conditions, to simulate the abusive treatment, affected the stems to such an extent that by the time they reached retail conditions, most of the stems were already brown.

Table 12. Supply chain scenario D (2018): changes in quality attributes of color (L, a* and b*), firmness (N), TSS (Brix °), TA (%), and stem browning index for 'Crimson Seedless' table grapes at harvest after 28 days cold storage (−0.3 ± 0.7 °C, 81.3 ± 4.1% RH), after 2 days 'abusive' temperature and humidity (25.1 ± 1.3 °C and 46.6 ± 6.0%RH), after another 10 days at retail conditions (5.4 ± 0.6 °C, 83.7 ± 2.9% RH) and then for 3, 7 and 10 days at ambient conditions (25.1 ± 1.3 °C and 46.6 ± 6.0%RH). Mean values with different letter(s) in the same column indicate statistically significant differences ($p < 0.05$).

				2018			
Time	**L***	**a***	**b***	**Firmness (N)**	**TSS (Brix °)**	**TA (%)**	**Stem Browning Index**
Harvest	27.46 ^{b}	7.91 ^{a}	9.36 ^{a}	116.60 ^{a}	17.80 ^{d}	0.79 ^{a}	1 ^{e}
28 days (−0.5 °C)	27.49 ^{b}	7.89 ^{a}	9.16 ^{a}	118.54 ^{a}	18.54 ^{c}	0.76 ^{b}	1.89 ^{e}
2 days (ambient)	28.21 ^{a}	7.65 ^{a}	9.23 ^{a}	115.72 ^{a}	17.75 ^{d}	0.75 ^{b}	3.25^{d}
10 days (5 °C)	27.71 ^{b}	7.88 ^{a}	9.54 ^{a}	123.01 ^{a}	19.55 ^{ab}	0.72 ^{c}	4.01 ^{c}
3 days	27.60 ^{b}	8.02 ^{a}	8.72 ^{a}	121.96 ^{a}	17.27 ^{d}	0.72 ^{c}	4.62 ^{b}
7 days	28.23 ^{a}	7.69 ^{a}	8.34 ^{a}	116.28 ^{a}	19.11 ^{b}	0.75 ^{b}	4.94 ^{a}
10 days	27.82 ^{b}	7.83 ^{a}	8.12 ^{a}	123.24 ^{a}	19.74 ^{a}	0.76 ^{b}	4.94 ^{a}
p-value	<0.01	0.86	0.21	0.23	<0.01	<0.01	<0.01

Note: Mean values within the same column with different letters are significantly ($p < 0.05$) different by Duncan's Multiple Range test (DMRT).

3.4. Socioeconomic Impacts of Postharvest Losses

Based on the percentage losses along the simulated supply chains, estimates were made to determine the volume of table grapes that could be lost at the national level (Table 13). In 2017, South Africa produced approximately 325,061 tons, of which 20,046 tons were sold locally, and 305,015 tons were exported [17]. The ranges provided in the following data are estimates made from the lowest losses which were recorded in the 2018 season to the highest losses that were recorded in 2017. It thus provides a range of losses that could occur in any given season.

Supply chain scenario A (marketing at ambient temperatures and relative humidity)

Losses translated were between 555 and 1052 tons after 3 days. This equates to a financial loss of R7.3 million–R13.8 million, 3,623,595–6,868,508 MJ of energy, 116,744–221,288 m^3 water used in production and 505–957 tons CO_2eq emissions. After 7 days, the losses increase to 575–1904 tons, R7.5 million–R25 million, 375,417–12,431,216 MJ, 120,951–400,466 m^3 and 523–1733 tons CO_2eq. After 10 days, 1305–2068 tons were lost, R17.1 million–R27.1 million, 8,520,345–13,461,972 MJ of energy, 274,507–435,003 m^3 and 1188–1882 tonnes CO_2eq.

Table 13. Impact of postharvest losses in terms of magnitude, monetary value, energy used, water footprint, and greenhouse gas emissions in the production and distribution of table grapes along different supply chains.

Year	Supply Chain Scenario	Storage Condition		Estimated Physical and Economic Losses		* Estimated Environmental and Resource Impacts		
		Time	Temp (°C) and Humidity (%)	Physical (ton)	Value (ZAR)	Energy (MJ)	Water Footprint (m^3)	Emissions CO_2eq (ton)
2017	A	3 days	25.1 ± 1.3 °C; 46.6 ± 6.0%RH	1052 ^{a}	13,816,968 ^{a}	6,868,508 ^{a}	221,288 ^{a}	957 ^{a}
2017	A	7 days	25.1 ± 1.3 °C; 46.6 ± 6.0%RH	1904 ^{b}	25,007,316 ^{b}	12,431,216 ^{b}	400,466 ^{b}	1733 ^{b}
2017	A	10 days	25.1 ± 1.3 °C; 46.6 ± 6.0%RH	2068 ^{b}	27,161,112 ^{b}	13,461,972 ^{b}	435,003 ^{b}	1882 ^{b}
2017	B	14 days	−0.3 ± 0.7 °C; 81.3 ± 4.1%RH	493 ^{a}	6,475,062 ^{a}	3,218,797 ^{a}	103,703 ^{a}	449 ^{a}
2017	B	10 days	5.4 ± 0.6 °C; 83.7 ± 2.9%RH	874 ^{ab}	11,479,116 ^{ab}	5,706,346 ^{ab}	183,846 ^{ab}	795 ^{ab}
2017	B	3 days	25.1 ± 1.3 °C; 46.6 ± 6.0%RH	1431 ^{bc}	18,794,754 ^{bc}	9,342,999 ^{bc}	301,011 ^{bc}	1302 ^{bc}
2017	B	7 days	25.1 ± 1.3 °C; 46.6 ± 6.0%RH	2074 ^{cd}	27,239,916 ^{cd}	13,541,146 ^{cd}	436,266 ^{cd}	1887 ^{cd}
2017	B	10 days	25.1 ± 1.3 °C; 46.6 ± 6.0%RH	2534 ^{d}	33,281,556 ^{d}	16,544,486 ^{d}	533,027^{d}	2306 ^{d}
2017	C	28 days	−0.3 ± 0.7 °C; 81.3 ± 4.1%RH	13,299 ^{a}	279,305,598 ^{a}	86,829,171 ^{a}	2,797,445 ^{a}	12,102 ^{a}
2017	C	10 days	5.4 ± 0.6 °C; 83.7 ± 2.9%RH	21,443 ^{b}	450,345,886 ^{b}	140,001,347 ^{b}	4,510,535 ^{b}	19,513 ^{b}
2017	C	3 days	25.1 ± 1.3 °C; 46.6 ± 6.0%RH	22,602 ^{b}	474,687,204 ^{b}	147,568,458 ^{b}	4,754,331 ^{b}	20,568 ^{b}
2017	C	7 days	25.1 ± 1.3 °C; 46.6 ± 6.0%RH	29,251 ^{b}	614,329,502 ^{b}	190,979,779 ^{b}	6,152,948 ^{b}	26,618 ^{b}
2017	C	10 days	25.1 ± 1.3 °C; 46.6 ± 6.0%RH	43,587 ^{c}	915,414,174 ^{c}	284,579,523 ^{c}	9,168,525 ^{c}	39,664 ^{c}
2018	A	3 days	25.1 ± 1.3 °C; 46.6 ± 6.0%RH	555 ^{a}	7,289,370 ^{a}	3,623,595^{a}	116,744 ^{a}	505,05 ^{a}
2018	A	7 days	25.1 ± 1.3 °C; 46.6 ± 6.0%RH	575 ^{a}	7,552,050 ^{a}	3,754,175^{a}	120,951 ^{a}	523,25 ^{a}
2018	A	10 days	25.1 ± 1.3 °C; 46.6 ± 6.0%RH	1305 ^{b}	17,139,870 ^{b}	8,520,345^{b}	274,507 ^{b}	118,755 ^{b}
2018	B	14 days	−0.3 ± 0.7 °C; 81.3 ± 4.1%RH	611 ^{a}	8,024,874 ^{a}	3,989,219 ^{a}	128,524 ^{a}	556 ^{a}
2018	B	10 days	5.4 ± 0.6 °C; 83.7 ± 2.9%RH	728 ^{ab}	9,561,552 ^{ab}	4,753,112 ^{ab}	153,135 ^{ab}	663 ^{ab}
2018	B	3 days	25.1 ± 1.3 °C; 46.6 ± 6.0%RH	1163 ^{bc}	15,274,842 ^{bc}	7,593,227 ^{bc}	244,637 ^{bc}	1058 ^{bc}
2018	B	7 days	25.1 ± 1.3 °C; 46.6 ± 6.0%RH	1303 ^{cd}	17,113,602 ^{cd}	85,07,287 ^{cd}	274,086 ^{cd}	1186 ^{cd}
2018	B	10 days	25.1 ± 1.3 °C; 46.6 ± 6.0%RH	1856 ^{d}	24,376,704 ^{d}	12,117,824 ^{d}	390,410 ^{d}	1689 ^{d}
2018	C	28 days	−0.3 ± 0.7 °C; 81.3 ± 4.1%RH	29,861 ^{a}	627,140,722 ^{a}	194,962,469 ^{a}	6,281,261 ^{a}	27,174 ^{a}
2018	C	10 days	5.4 ± 0.6 °C; 83.7 ± 2.9%RH	11,560 ^{b}	242,783,120 ^{b}	75,475,240 ^{b}	2,431,646 ^{b}	10,520 ^{b}
2018	C	3 days	25.1 ± 1.3 °C; 46.6 ± 6.0%RH	12,109 ^{b}	254,313,218 ^{b}	79,059,661 ^{b}	2,547,128 ^{b}	11,019 ^{b}
2018	C	7 days	25.1 ± 1.3 °C; 46.6 ± 6.0%RH	21,107 ^{ab}	443,289,214 ^{ab}	137,807,603 ^{ab}	4,439,857 ^{ab}	19,207 ^{ab}
2018	C	10 days	25.1 ± 1.3 °C; 46.6 ± 6.0%RH	31,173 ^{a}	654,695,346 ^{a}	203,528,517 ^{a}	6,557,241 ^{a}	2,836,743 ^{a}
2018	D	28 days	−0.3 ± 0.7 °C; 81.3 ± 4.1%RH	9242 ^{a}	194,100,484 ^{a}	60,341,018 ^{a}	1,944,055 ^{a}	8410 ^{a}
2018	D	2 days	25.1 ± 1.3 °C; 46.6 ± 6.0%RH	9730 ^{a}	204,349,460 ^{a}	63,527,170 ^{a}	2,046,706 ^{a}	8854^{a}
2018	D	10 days	5.4 ± 0.6 °C; 83.7 ± 2.9%RH	13,024 ^{ab}	273,530,048 ^{ab}	85,033,696 ^{ab}	2,739,598 ^{ab}	11,852 ^{ab}
2018	D	3 days	25.1 ± 1.3 °C; 46.6 ± 6.0%RH	15,251 ^{bc}	320,301,502 ^{bc}	99,573,779 ^{bc}	3,208,048 ^{bc}	13,878 ^{bc}
2018	D	7 days	25.1 ± 1.3 °C; 46.6 ± 6.0%RH	20,314 ^{c}	426,634,628 ^{c}	132,630,106 ^{c}	4,273,050 ^{c}	18,486 ^{c}
2018	D	10 days	25.1 ± 1.3 °C; 46.6 ± 6.0%RH	27,177 ^{d}	570,771,354 ^{d}	177,438,633 ^{d}	5,716,682 ^{d}	24,731 ^{d}

Note: a,b,c Values in a column without a common superscript are significantly different ($p < 0.05$). * Estimated values obtained using the volume of table grapes sold locally, 20,046 t and exported, 305,015 t [17]. This section may be divided by subheadings. It should provide a concise and precise description of the experimental results, their interpretation as well as the experimental conclusions that can be drawn.

Supply chain scenario B (to local retail markets)

After 14 days in cold storage, the losses were between 493 and 611 tons with a financial loss of R6.5 million–R8 million, lost energy of 3,218,797–3,989,219 MJ, a water footprint of 103,703–128,524 m^3, and 449–556 tons CO_2eq. After 10 days at retail conditions, losses were 728–874 tons, R9.6 million–R11.5 million, 4,753,112–5,706,346 MJ, 153,135–183,846 m^3 water lost, and 663–795 tons CO_2eq.

Once moved to ambient conditions, the losses after 3 days were 1163–1431 tons, R15.2 million–R18.8 million, 7,593,227–9,342,999 MJ, 244,637–301,011 m^3, and 1058–1302 tons CO_2eq. After 7 days, 1303–2074 tons, R17.1 million–R27.2 million, 8,507,287–13,541,146 MJ, 274,086–436,266 m^3, and 1186–1887 tons CO2eq. After 10 days, 1856–2534 tons, R24.3 million–R33.3 million, 12,117,824–16,544,486 MJ, 390,410–533,027 m^3, and 1689–2306 tons CO_2eq.

Supply chain scenario C (to export retail markets)

After 28 days in cold storage, the losses were between 13,299 and 29,861 tons with a financial loss of R279.3 million–R627.1 million, 86,829,171–194,962,469 MJ of energy, 2,797,445–6,281,261 m^3 of water, and 12,102–27,174 tons CO_2eq. After 10 days at retail conditions, 11,560–21,443 tons, R242.8 million–R450.3 million, 75,475,240–140,001,347 MJ, 2,431,646–4,510,535 m^3, and 10,520–19,513 tons CO_2eq. After 3 days at ambient conditions, losses were 12,109–22,602 tons, R254.3 million–R474.7 million, 79,059,661–147,568,458 MJ, 2,547,128–4,754,331 m^3 and 11,019–20,568 tons CO_2eq. After 7 days: 21,107–29,251 tons, R443.3 million–R614.3 million, 137,807,603–190,979,779 MJ, 4,439,857–6,152,948 m^3, and 19,207–26,618 tons CO_2eq. After 10 days at ambient, the losses were 31,173–43,587 tons, R654.7 million–R915.4 million, 203,528,517–284,579,523 MJ, 6,557,241–9,168,525 m^3, and 28,367–39,664 tons CO_2eq.

Supply chain scenario D (simulated 'abusive' treatment of fruit within the export chain)

After 28 days in cold storage, the losses were 9242 tons with a financial value of R194.1 million, 60,341,018 MJ of energy, 1,944,055 m^3 water, and 8410 tons CO_2eq. After 2 days at 'abusive' ambient temperature and humidity before entering retail conditions, the losses were 9730 tons, R204.3 million, 63,527,170 MJ, 2,046,706 m^3 water, and 8854 tons CO_2eq. After 10 days at retail conditions, losses were 13,024 tons, R273.5 million, 85,033,696 MJ, 2,739,598 m^3, and 11,852 tons CO_2eq. At 3 days of ambient conditions, losses were 15,251 tons, R320.3 million, 99,573,779 MJ, 3,208,048 m^3, and 13,878 tons CO_2eq. After 7 days ambient: 20,314 tons, R426.6 million, 132,630,106 MJ, 4,273,050 m^3 and 18,486 tons CO_2eq. After 10 days at ambient conditions, the losses were 27,177 tons, R570.8 million, 177,438,633 MJ, 5,716,682 m^3, and 24,731 tons CO_2eq.

4. Discussion

4.1. Physical Losses at Farm Level

The losses measured on farm level in 2017 were higher than the findings of [35], who concluded in an economic analysis of the South African table grape supply chain that approximately 9.5% was lost between farm and intake. However, the data used in that study was based on perception data gathered from different role-players in the table grape industry, the authors own elaborations and from [18] as no primary data was collected.

The authors of [36] reported similar losses at agricultural production level of 15%, where table grape losses were quantified along the supply chain in Iran. The materials and methods of that study are unclear, however, as it divides the supply chain into production, postharvest, processing, distribution, and consumption stages without clearly describing the various stages. In the present study, the losses at farm level include what it seems they refer to as production, postharvest and processing into one. If that is the case, the losses experienced in Iran were much higher than our findings and amounted to 46% of total production. In [36], the authors used data from government and private sources with estimates and interviews, and no primary data was collected in that study.

In the present study, the average loss measured on farm level in 2018 was 5.97%. This is less than reported by [35,36] but very similar to the findings of [37] that reported losses of table grapes in Pakistan of 4.8% at farm level and [38] where losses of table grapes at farm level in India were reported as 3.4% for grapes prepared for the domestic market while grapes prepared for the export market sustained losses of 7.82% as the requirements for export produce are stricter.

The losses measured in 2017 are similar to that of [38] for export grapes (Thomson Seedless), but the reasons for the losses differ, water berries (6.72%), harvest injury (0.57%), mummies (0.02%), and immature (0.26%). The authors of [39] report losses of table grape (Nana Purple) on farm level as 8–10% due to insufficient coloring, while in [36,37] the authors do not specify the reasons for losses.

The large differences between farms in 2017, were unexpected and although it was initially thought that it could have been due to different climatic conditions or soil types it appears to have been related to different vineyard management practices as there was no statistical difference between losses on the farms in the 2018 season and the average losses were less than half recorded in 2017. The reasons for this could be two-fold. Firstly, by having been made aware of how high the losses were during the previous season, the farm managers took steps to reduce this during the 2018 season, workers were trained to be more careful when handling the crates after harvest. Secondly, the farm that sustained the highest losses (23.3%) in 2017 harvested later than was optimal and therefore the bunches stayed on the vines too long. The 2018 harvest occurred two weeks earlier than in 2017, and the grapes were in better condition leading to fewer losses on farm level.

4.2. Physical Losses along the Simulated Supply Chain

Supply chain scenario A (handling and marketing fruit under ambient conditions)

These results support the research of [40] that reported the mass of grapes cv. 'Crimson Seedless' always decreased with time at all combinations of storage temperature and RH. Grapes stored at higher temperature lost weight faster than those maintained at lower temperatures. The increased vapor pressure deficit and respiration rates of the stems of grapes stored at higher temperatures accelerated transpiration rates of fruit. In [19], the authors further found a linear profile for the mass decreasing in grapes of cv. 'Thompson' and 'Superior'. Similar findings were also reported by [39], noting a 5% weight loss in grapes transported in trucks in temperatures of 35–40 °C before reaching the wholesale market.

The findings on the incidence of decay were similar to the findings of [20], who found a severe incidence of decay (2–5 infected berries per carton) of 'Regal Seedless' table grapes after 7 days shelf life at 24.33 ± 0.04 °C in similar packaging to that used in this study. The authors of [41] reported much higher levels of 40.5% after 7 days at 15 °C for cv 'Thompson Seedless'. Visible decay in a carton could make the carton hard to sell, even though less than 10% of the berries are affected, as from the consumer's perspective, appearance is the first factor that influences purchase decision, followed by perceived value for money and fruit eating quality [42]. This is similar to the findings of [20] who found severe incidence of decay (2–5 infected berries per carton) of 'Regal Seedless' table grapes after 7 days shelf life at 24.33 ± 0.04 °C in similar packaging to that used in this study. The authors of [41] reported much higher levels of 40.5% after 7 days at 15 °C for cv 'Thompson Seedless'. Visible decay in a carton could make the carton hard to sell, even though less than 10% of the berries are affected, as from the consumer's perspective, appearance is the first factor that influences purchase decision, followed by perceived value for money and fruit eating quality [42].

Supply chain scenario B (to local retail markets)

After 10 days at retail conditions, the reported decay was less than half the amount of 4.56% reported by [38]. Similar to [43] reporting 1% SO_2 damage for cv 'Red Globe' after 15 days in cold storage, the incidence of SO_2 berry damage in this study was 1% after 14 days in cold storage. The authors of [20,44] reported that the combination of free water (100% RH), as occurs with the formation of condensation when cartons are removed from cooler conditions to ambient conditions, combined with SO_2 in nonperforated liners may result in the formation of acidic conditions that may increase SO_2 injury, this seems to be the case in this study also. It is suggested that after a few days at ambient conditions, the free water evaporates, and the damage stops. Based on the investigated seasons, the decreases of 5.4% and 6.8% in weight noted for the two seasons were more than double the amount of 2% weight loss after 14 days of cold storage for cv. 'Thompson Seedless' reported by [45], while findings on decay during cold storage were similar to the 0% decay reported.

Supply chain scenario C (to export retail markets)

The measured weight loss was similar to the 5% after 10 days and 10% after 14 days at room temperature (25 °C) and 45–70% relative humidity reported by [46] for cv 'Victoria'. After 28 days in cold storage (−0.3 ± 0.7 °C, 81.3 ± 4.1% RH), there was a 2.14% incidence of decay in 2017 and 0.94% in 2018. These results were similar the mean of 1.28% decay reported by [47] for cultivars 'Red Globe', 'Sunred Seedless' and 'Thompson Seedless' under similar conditions for the same time period. After 10 days at retail display conditions

(5.4 ± 0.6 °C, 83.7 ± 2.9% RH) this increased to 3.2% in 2017 and 2.6% in 2018. These results were slightly higher than the 2.5% reported by [35] for the perceived losses of table grapes at retail level and correspond to the lowest end of the 3–7% range of loss reported for fresh fruit under retail conditions in the UK and Spain by [48]. Both [35,48], however, used data collected through interviews with managers in food manufacturing and questionnaires completed by other role players in the table grape supply chain while no primary sampling data was collected. Three days after being moved to ambient temperature and humidity conditions (25.1 ± 1.3 °C and 46.6 ± 6.0%RH), decay increased to 4.44% in 2017 and 3.16% in 2018. The authors of [49] reported decay of more than a hundred berries per kg after 4 weeks at 0 °C and 3 days shelf-life at 20 °C for cv. 'Thompson Seedless'. Taking the average weight of a 'Thompson Seedless' berry as 5g reported [50] the data translates to 500g infected berries per kg or 50%. That is much higher than the decay rate measured in this study. The results for SO_2 damage seem similar to the rating of 4 (11–20 berries per replicate consisting of 10 bunches) with SO_2 damage after 65 days at 0 °C and 3 days at 20 °C reported by [51] although it is not easy to compare as that study used a rating of 1–5 for describing SO_2 damage and not % and it is unknown exactly how many berries were in a replicate of 10 bunches.

Supply chain scenario D (simulated 'abusive' treatment of fruit within the export chain)

The 5.7% decrease in weight noted is half the weight loss of around 12% reported by [52] for 'Crimson Seedless' table grapes under similar conditions and for the same time period.

Decay and $S0_2$ damage are disorders that can be caused or aggravated by wet berries in combination with elevated temperature [53]. Results indicate that the decay in this trial was lower than that recorded for supply chain scenario C, which was the same in all regards except for the 2 days under ambient conditions. This could be due to condensation evaporating, leaving less free moisture that would exacerbate decay.

Total Amount of Physical Losses

Supply chain scenario A (marketing at ambient conditions)

For both years, the losses were considerably less than the 53% reported by [36] for table grape losses along the supply chain under ambient conditions in Iran. No sampling data was recorded in that study. However, the data used for their calculations were collected through government and private data sources with horticulture expert estimates, grape grower interviews, agriculture cooperation interviews, and market consultations.

Supply chain scenario B (to local retail markets)

When these losses are taken only from harvest to retail level, it translates to 18.26% loss in 2017 and 9.62% in 2018. The result for 2018 is similar to [38], who reported losses of 7.96% from the field to retailer in India.

Supply chain scenario C (to export retail markets)

The 2018 data was similar to [38] reporting export supply chain losses for table grapes in India, of 19.95%, as well as the approximate figure of 15.5% reported by [35] for the South African table grape supply chain, although no sampling data was collected for those studies. The 2017 data of this study was significantly higher and showed how variable the yearly losses could be.

Supply chain scenario D (simulated 'abusive' treatment of fruit within the export chain)

In terms of quantity of losses, therefore, the 2 days at ambient conditions in the middle of the cold chain did not cause a significant difference. It did, however, create a difference in quality, as will be illustrated in the next section.

4.3. Quality Losses along the Supply Chain

Supply chain scenario A (marketing at ambient conditions)

Results for firmness over time supports the findings of [23] reporting no difference in firmness for cv. 'Mystery' after 6 days under ambient conditions (22–28 °C) but contrasts with the findings of [54] for cv. 'Muscat Hamburg' reporting a significant decrease of firmness over time under ambient conditions. Similar findings of unchanging TSS and TA values were reported for both cv. Müşküle and cv. Red Globe by [27] during the first week of that research project. The results on change in stem color support the report by [19]. The author reported major rachis browning during marketing at ambient temperatures and relative humidity.

Supply chain scenario B (to local retail markets)

The increase in lightness measured in 2018 differ to the findings of [45] that found the L values for the black table grape cv. 'Alphonse Lavallée' decrease during cold storage, indicating that the grapes became darker over time. Berry firmness results that showed no significant difference over time is similar to findings of [55] reporting no significant difference in firmness for cv. 'Italia' after cold storage and 7 days shelf life. The increase in TA measured during the 2018 season is similar [27] also describing such a significant increase in TA after 2 and 3 weeks in storage for cv. Müşküle and cv. Red Globe. The authors of [23,56] describe increases in TSS of grapes during cold storage due to water loss. In contrast to the 2018 findings, yet similar to the 2017 findings of this study, TSS levels could remain stable under different storage conditions and for different cultivars. In [24] the authors reported that TSS levels in cv. Red Globe, after up to 12 weeks in storage under different controlled atmosphere (CA) conditions at 0 °C, was almost equal to that measured at harvest.

Supply chain scenario C (to export retail markets)

No significant difference in any of the color attributes were found in 2017, similar to findings of [57] for cv. 'BRS Isis' after 50 days in cold storage plus 5 days under ambient conditions. In 2018, however, results indicated that berry color became lighter, which differs from [28] reporting that L decreased with storage time for cv. 'Thompson Seedless', while [55] reported no significant changes in color for cv. 'Italia' after 50 days of cold storage and 7 days under ambient conditions.

No change in firmness was observed. Similarly, the authors of [55] reported no significant difference in firmness for cv. Italia after up to 50 days in cold storage and 7 days shelf life. No changes were observed in TSS, similar to findings of [43] for cv. 'Red Globe'. Results for changes in TA are similar to findings reported by [28] that berry TA underwent a progressive decrease during storage for cv. 'Thompson Seedless'. Stem color changes support the findings of [21,53] also noted for cv. 'Thompson Seedless' that the average stem condition deteriorated more during the shelf life period than cold storage.

Supply chain scenario D (simulated 'abusive' treatment of fruit within the export chain).

Unchanged berry firmness was also reported by [23] for cv. 'Mystery' and [43] for cv. 'Red Globe'. [19], however, reported a decrease in firmness during shelf life trials for the white grape cultivars 'Superior' and 'Thompson' held in high (>95%) and low (70%) relative humidity at 20 or 10 °C for up to 11 d. Several researchers similarly reported general increases in TSS of grapes during cold storage, including [23,56]. This is attributed to the gluconeogenesis pathway or water loss. The authors of [23] also reported similar findings on stem browning for cv. 'Mystery' where cooling delays of 48 h resulted in a rachis browning score of 5, indicating very severe damage.

4.4. Socioeconomic Impacts of Postharvest Losses

The socioeconomic impacts of these postharvest losses indicate a financial loss of between R279 million and over R600 million annually for the table grape export industry. The authors of [35] estimated losses of approximately 9.5% between production and intake stages for the South African table grape industry, translating into a financial loss of R270.5 million with an additional 2.2% or R93.2 million between intake and export and 3.8% or R400,000 between the importer to retail depot. It is unclear how 3.8% equals R400,000 when it was also stated that 2.2% equals R93,200,000, the accuracy of the estimates are therefore uncertain. The values given for losses between production and intake does, however, approximate the values this current study recorded for the 2018 season, while the losses during the 2017 season were more than double that amount.

Additionally, as much as 177.43 million MJ of fossil energy and 4.8 million m^3 of freshwater resources were lost. At the Eskom tariff rate of R0.90 per kWh, the lost energy is worth R44.36 million [58]. The fresh water lost could sustain at least 263,013 individuals daily for a whole year at daily minimum usage rate of 0.05 m^3 per day. Losses also contribute to unwanted emission of approximately 52,263 tons of CO_2eq, contributing to environmental degradation from greenhouse gases.

5. Conclusions

The highest loss in the supply chain was measured at the farm level. It is therefore important to include this stage when studies are conducted on the quantification of postharvest losses. As the main reason for losses at this stage was mechanical damage due to the rough handling of bunches and crates causing berries to drop off the bunches as well as the crushing of berries due to loading too many bunches in crates, these losses could be improved by making workers more aware of the necessity to handle crates with care. The harvest timing is also essential, as delayed harvesting reduces shelf life and results in an increased postharvest loss.

The main quality problem, among all supply chain scenarios, was rachis and stem browning at temperatures higher than −0.5 °C. This caused berries to drop faster and bunches to look less fresh, as well as causing bunches to weigh less when sold. While 500 g or 1 kg punnets are routinely kept at around 5 °C at the retail level, during peak season 4.5–10 kg cartons are often stacked on the floor under ambient conditions. Therefore, the table grapes would have a maximum shelf-life of 7 days before the stems have browned, and too many berries per bunch are decayed to sell. It would be advisable to keep cartons at −5 °C and high RH and only place bunches in punnets in 5 °C display fridges as the stock sells.

The increase in weight loss and especially stem browning recorded in Scenario D ('abusive' treatment of fruit within the export chain), compared to Scenario C (shipping to export markets) indicated the importance of eliminating the delay between reefer delivery and quality checking as a break in the cold chain of 2 days has a significant impact on the quality of the bunches and therefore also the price it can be sold at.

This study was conducted on farms with good infrastructure, cultivation practices and cooling facilities, where nonetheless, farm-level losses of up to 23% were recorded. It is significant that during preseason interviews with farm management, the highest estimate of losses was 13%, most of them lower. As 'Crimson Seedless' is a high value crop, even relatively small improvements in future could have a large financial impact for producer-exporters.

In the changing local agricultural environment of many more upcoming farmers entering the industry, this situation deserves much more attention than what was the case so far. On a global level, as we are approaching population levels of around nine billion people, the choice is obvious: not only must we produce more food, but we should also waste much less.

Author Contributions: Conceptualization, U.L.O.; Formal analysis, A.B.; Funding acquisition, U.L.O.; Investigation, A.B.; Methodology, A.B., U.L.O., O.A.F.; Supervision, O.A.F. and U.L.O.; Validation, O.A.F. and U.L.O.; Visualization, A.B.; Writing—original draft, A.B.; Writing—review and editing, A.B., U.L.O., O.A.F. All authors have read and agreed to the published version of the manuscript.

Funding: This work is based on the research supported wholly/in part by the National Research Foundation of South Africa (Grant Numbers: 64813).

Institutional Review Board Statement: Not applicable.

Informed Consent Statement: Not applicable.

Data Availability Statement: Design of the study; in the collection, analyses, or interpretation of data; in the writing of the manuscript, or in the decision to publish the results. The opinions, findings and conclusions or recommendations expressed are those of the author(s) alone, and the NRF accepts no liability whatsoever in this regard.

Acknowledgments: The authors offer their heartfelt thanks to the Karsten Group.

Conflicts of Interest: The authors declare no conflict of interest.

References

1. Kuhr, R.J. Integrated pest management. A new strategy in an old war. *N. Y. Food Life Sci.* **1979**, *12*, 3.
2. FAO. *State of Food Insecurity in the World: Economic Crises—Impacts and Lessons Learned*; Food and Agriculture Organization of the United Nations: Rome, Italy, 2009. Available online: https://www.fao.org (accessed on 12 September 2020).
3. Hodges, R.J.; Buzby, J.C.; Bennett, B. Postharvest losses and waste in developed and less developed countries: Opportunities to improve resource use. *J. Agric. Sci.* **2010**, *149*, 37–45. [CrossRef]
4. Alexandratos, N. *How to Feed the World in 2050*; Expert meeting; FAO Headquarters: Rome, Italy, 2009.
5. FAO. *Food Losses and Waste in Europe and Central Asia*; Food and Agriculture Organization of the United Nations: Rome, Italy, 2013. Available online: https://www.fao.org (accessed on 11 November 2020).
6. Statista. Global Volume of Wasted Food as of 2017, by Food Category, Global Waste Generation Dossier, Statista. 2019. Available online: https://www.statista.com/study/58790/global-waste-generation/ (accessed on 10 September 2020).
7. Smil, V. Improving Efficiency and Reducing Waste in Our Food System. *Environ. Sci.* **2004**, *1*, 17–26. [CrossRef]
8. Parfitt, J.; Barthel, M.; Macnaughton, S. Food waste within food supply chains: Quantification and potential for change to 2050. *Phil. Trans. R. Soc.* **2010**, *365*, 3065–3081. [CrossRef] [PubMed]
9. FLW Protocol (Food Loss and Waste Protocol). *The Food Loss and Waste Accounting and Reporting Standard*; World Resources Institute: Washington, DC, USA, 2016.
10. Lipinski, B.; Hanson, C.; Lomaw, J.; Kitinoja, L.; Waite, R.; Searchinger, T. *Reducing Food Loss and Waste*; Working Paper; Installment 2 of Creating a Sustainable Food Future; World Resources Institute: Washington, DC, USA, 2013. Available online: http://www.worldresourcesreport.org (accessed on 23 December 2020).
11. Kader, A.A. Increasing availability by reducing post-harvest losses of fresh produce. *Acta Hortic.* **2005**, *682*, 2169–2175. [CrossRef]
12. Munhuweyi, K.; Opara, U.L.; Sigge, G. Postharvest losses of cabbages from retail to consumer and the socio-economic and environmental impacts. *Br. Food J.* **2016**, *118*, 286–300. [CrossRef]
13. Affognon, H.; Mutungi, C.; Sanginga, P.; Borgemeister, C. Unpacking postharvest losses in Sub-Saharan Africa: A meta-analysis. *World Dev.* **2015**, *66*, 49–68. [CrossRef]
14. Oelofse, S.H.H.; Nahman, A. Estimating the magnitude of food waste generated in South Africa. *Waste Manag. Res.* **2013**, *31*, 80–86. [CrossRef] [PubMed]
15. Kitinoja, L.; Kader, A.A. *Measuring Post-Harvest Losses of Fresh Fruits and Vegetables in Developing Countries*; White paper 15-02; Post-harvest Education Foundation: La Pine, OR, USA, 2015.
16. Gustavsson, J.; Cederberg, C.; Sonesson, U.; Van Otterdijk, R.; Meybeck, A. *Global Food Losses and Food Waste Section 3.2*; Study Conducted for the International Congress "Save Food!" at Interpack2011, Düsseldorf, Germany; FAO Rural Infrastructure and Agro-Industries Division 38: Rome, Italy, 2011.
17. DAFF. A Profile of the South African Table Grapes Market Value Chain. 2017. Available online: http://webapps.daff.gov.za (accessed on 31 August 2019).
18. SATI. South African Table Grape Industry Statistics Booklet. 2016. Available online: http://www.satgi.co.za/Media/Default/Documents/Booklets/SATI_BOOK_2016.pdf (accessed on 19 August 2020).
19. Lichter, A.; Kaplunov, T.; Zutahy, Y.; Daus, A.; Alchanatis, V.; Ostrovsky, V.; Lurie, S. Physical and visual properties of grape rachis as affected by water vapour pressure deficit. *Postharvest Biol. Technol.* **2011**, *59*, 25–33. [CrossRef]
20. Ngcobo, M.E.K.; Opara, U.L.; Thiart, G.D. Effects of packaging liners on cooling rate and quality attributes of table grape (cv Regal Seedless). *Packag. Technol. Sci.* **2012**, *25*, 73–84. [CrossRef]
21. Ngcobo, M.E.K.; Delele, M.A.; Opara, U.L.; Meyer, C.J. Performance of multipackaging for table grapes based on airflow, cooling rates and fruit quality. *J. Food Eng.* **2013**, *116*, 613–621. [CrossRef]
22. Crisosto, C.H.; Smilanick, J.L.; Dokoozlian, N.K. Table grapes suffer water loss, stem browning during cooling delays. *Calif. Agric.* **2001**, *55*, 39–42. [CrossRef]
23. Li, L.; Kaplunov, T.; Zutahy, Y.; Daus, A.; Porat, R.; Lichter, A. The effects of 1-methylcyclopropane and ethylene on postharvest rachis browning in table grapes. *Postharvest Biol. Technol.* **2015**, *107*, 16–22. [CrossRef]
24. Crisosto, C.H.; Garner, D.; Crisosto, G. Carbon dioxide-enriched atmospheres during cold storage limit losses from Botrytis but accelerate rachis browning of 'Red Globe' table grapes. *Postharvest Biol. Technol.* **2002**, *26*, 181–189. [CrossRef]
25. Jiang, L.; Jin, P.; Wang, L.; Xuan, Y.; Wang, H.; Zheng, Y. Methyl jasmonate primes defense responses against Botrytis cinerea and reduces disease development in harvested table grapes. *Sci. Hortic.* **2015**, *192*, 218–223. [CrossRef]
26. Ngcobo, M.E.K.; Delele, M.A.; Pathare, P.B.; Chan, L.; Opara, U.L.; Meyer, C.J. Moisture loss characteristics of fresh table grapes packed in different film liners during cold storage. *Biosyst. Eng.* **2012**, *113*, 363–370. [CrossRef]

27. Sabir, F.K.; Sabir, A. Quality response of table grapes (*Vitis vinifera* L.) during cold storage to postharvest cap stem excision and hot water treatments. *Int. J. Food Sci. Technol.* **2013**, *48*, 999–1006. [CrossRef]
28. Sabir, F.K.; Sabir, A. Extending postharvest quality attributes of grapes (*V. vinifera* L. cv. 'Thompson Seedless') by preharvest calcium pulverizations. *Acta Sci. Pol. Hortorum Cultus* **2017**, *16*, 29–38. [CrossRef]
29. Kitinoja, L.; Al Hassan, H.Y. Identification of appropriate postharvest technologies for small scale horticultural farmers and marketers in Sub-Saharan Africa and South Asia—Part 1. Postharvest losses and quality assessments. *Acta Hortic.* **2012**, *934*, 31–40. [CrossRef]
30. Pathare, P.B.; Opara, U.L.; Al-Said, F.A.-J. Colour measurement and analysis in fresh and processed foods: A review. *Food Bioprocess Technol.* **2013**, *6*, 36–60. [CrossRef]
31. Chen, L.; Opara, U.L. Texture measurement approaches in fresh and processed foods—A review. *Food Res. Int.* **2013**, *51*, 823–835. [CrossRef]
32. Janse van Vuuren, P.F. Regional Resource Flow Model: Fruit Sector Report. GreenCape. 2015. Available online: https://www.green-cape.co.za/assets/Bioeconomy/RRFM-Fruit-sector-report-2014-15.pdf (accessed on 18 October 2020).
33. Steenwerth, K.L.; Strong, E.B.; Greenhut, R.F.; Williams, L.; Kendall, A. Life cycle greenhouse gas, energy and water assessment of wine grape production in California. *Int. J. Life Cycle Assess.* **2015**, *20*, 1243–1253. [CrossRef]
34. Kangueehi, G.N. Water Footprint Analysis to Improve Water Use Efficiency in Table Grape (*Vitis vinifera* L. cv. Crimson Seedless) Production. A South African Case Study. Ph.D. Thesis, Department of Viticulture and Oenology, Faculty of AgriSciences, Stellenbosch University, Stellenbosch, South Africa, 2018.
35. Louw, L. Economic Aspects of Losses and Waste: Case Study of the South African Table Grape Supply Chain. Master' Thesis, Department of Agricultural Economics, Extension and Rural Development, Faculty of Natural and Agricultural Sciences, University of Pretoria, Pretoria, South Africa, 2017.
36. Rajabi, S.; Lashgarara, F.; Omidi, M.; Hosseini, S.J.F. Quantifying the grapes losses and waste in various stages of supply chain. *Biol. Forum* **2015**, *7*, 225–229.
37. Aujla, K.M.; Shah, N.A.; Ishaq, M.; Fraooq, A. Post-harvest losses and marketing of grapes in Pakistan. *Sarhad J. Agric.* **2011**, *27*, 485–490.
38. Murthy, M.R.K.; Reddy, G.P.; Rao, K.H. Retail marketing of fruits and vegetables in India: A case study on export of grapes from Andhra Pradesh, India. *Eur. J. Logist. Purch. Supply Chain Manag.* **2014**, *2*, 62–70.
39. Sharma, A.K.; Sawant, S.D.; Somkuwar, R.G.; Naik, S. *Postharvest Losses in Grapes: Present Indian Status*; Technical report ICAR; National Research Centre for Grapes: Pune, India, 2018. Available online: https://www.researchgate.net/publication/311693568 (accessed on 5 August 2020).
40. Pereira, E.; e Silva, R.G.B.; Spagnol, W.A.; Silveira Junior, V. Water loss in table grapes: Model development and validation under dynamic storage conditions. *Food Sci. Technol. Camp.* **2018**, *38*, 473–479. [CrossRef]
41. Mlikota Gabler, F.; Fassel, R.; Mercier, J.; Smilanick, J.L. Influence of temperature, inoculation interval, and dosage on biofumigation with Muscodor albus to control postharvest gray mold on grapes. *Plant Dis.* **2006**, *90*, 1019–1025. [CrossRef]
42. Eccher Zerbini, P. The quality of pear fruits. *Acta Hortic.* **2002**, *596*, 805–810. [CrossRef]
43. Sortino, G.; Allegra, A.; Passafiume, R.; Gianguzzi, G.; Gullo, G.; Gallotta, A. Postharvest Application of Sulphur Dioxide Fumigation to Improve Quality and Storage Ability of "Red Globe" Grape Cultivar During Long Cold Storage. *Chem. Eng. Trans.* **2017**, *58*, 403–408.
44. Zoffoli, F.P.; Latorre, B.A.; Naranjo, P. Hairline, a postharvest cracking disorder in table grapes induced by sulfur dioxide. *Postharvest Biol. Technol.* **2008**, *47*, 90–97. [CrossRef]
45. Sabir, F.K.; Sabir, A.; Unal, S.; Taytak, M.; Kucukbasmaci, A.; Bilgin, O.F. Postharvest quality extension of minimally processed table grapes by chitosan coating. *Int. J. Fruit Sci.* **2019**, *19*, 347–358. [CrossRef]
46. Xu, W.; Li, D.; Fu, Y.; Liu, Z.; Wang, Y.; Yu, X.; Shang, W. Chinese Packaging Institute—2012 Conference Extending the Shelf Life of Victoria Table Grapes by High Permeability and Fungicide Packaging at Room Temperature. *Packaging Technol. Sci.* **2013**, *26*, 43–50. [CrossRef]
47. Klaasen, J.A.; Van der Merwe, J.A.; Vries, F.A.; Calitz, F.J. Long-Term Storage Quality of Table Grapes as Influenced by Pre-Harvest Yeast Applications and Post-Harvest Use of Controlled Atmosphere. *S. Afr. J. Enol. Vitic.* **2006**, *27*, 2. [CrossRef]
48. Mena, C.; Adenso-Diaz, B.; Yurt, O. The causes of food waste in the supplier–retailer interface: Evidences from the UK and Spain. *Resour. Conserv. Recycl.* **2011**, *55*, 648–658. [CrossRef]
49. Lichter, A.; Zutkhy, Y.; Sonego, L.; Dvir, O.; Kaplunov, T.; Sarig, P.; Ben-Arie, R. Ethanol controls postharvest decay of table grapes. *Postharvest Biol. Technol.* **2002**, *24*, 301–308. [CrossRef]
50. Peacock, B.; Simpson, T. The Relationship between Berry Weight, Length and Width for Five Table Grape Varieties. Cooperative Extension Pub.-TBI 95. 2017. Available online: http://cetulare.ucanr.edu/files/82015.pdf (accessed on 10 July 2020).
51. Lichter, A.; Zutahy, Y.; Kaplunov, T.; Lurie, S. Evaluation of table grape storage in boxes with sulfurdioxide releasing pads with either an internal plastic liner or external wrap. *HortTechnology* **2008**, *18*, 206–214. [CrossRef]
52. Palou, L.; Serrano, M.; Martínez-Romero, D.; Valero, D. New approaches for quality retention of table grapes. *Fresh Prod.* **2010**, *4*, 103–110.
53. Du Plessis, S.F. Effects of Packaging and Postharvest Cooling on Quality of Table Grapes (*Vitis vinifera* L.). Master' Thesis, Department of Horticulture, Faculty of AgriSciences, Stellenbosch University, Stellenbosch, South Africa, 2003.

54. Tamizheezham, U.; Muthuvel, I.; Subbiah, A. Effect of Temperature on Shelf life of Muscat Hamburg Grapes under Storage. *Madras Agric. J.* **2018**, *105*, 426–429. [CrossRef]
55. Ahmed, S.; Roberto, S.R.; Domingues, A.R.; Shahab, M.; Junior, O.J.C.; Sumida, C.H.; De Souza, R.T. Effects of Different Sulfur Dioxide Pads on Botrytis Mold in 'Italia' Table Grapes under Cold Storage. *Horticulturae* **2018**, *4*, 29. [CrossRef]
56. Sabir, A.; Sabir, F.K.; Kara, Z. Effects of modified atmosphere packing and honey dip treatments on quality maintenance of minimally processed grape *cv.* Razaki (*V. vinifera* L.) during cold storage. *J. Food Sci Technol.* **2011**, *48*, 312–318. [CrossRef] [PubMed]
57. Colombo, R.C.; Carvalho, D.U.; Da Cruz, M.A.; Sumida, C.H.; Ahmed, S.; Bassoli, P.A.; De Souza, R.T.; Roberto, S.R. Cold Storage and Biocontrol Agents to Extend the Storage Period of 'BRS Isis' Seedless Table Grapes. *Horticulturae* **2018**, *4*, 18. [CrossRef]
58. Eskom. Available online: https://www.eskom.co.za/CustomerCare/TariffsAndCharges/Pages/Tariffs_And_Charges.aspx (accessed on 10 October 2019).

sustainability

MDPI

Article

Assessing the Camelina (*Camelina sativa* (L.) Crantz) Seed Harvesting Using a Combine Harvester: A Case-Study on the Assessment of Work Performance and Seed Loss

Walter Stefanoni [1], Francesco Latterini [1], Javier Prieto Ruiz [2], Simone Bergonzoli [3,*], Nadia Palmieri [1] and Luigi Pari [1]

1 Consiglio per la Ricerca in Agricoltura e l'Analisi dell'Economia Agraria (CREA), Centro di Ricerca Ingegneria e Trasformazioni Agroalimentari, Via della Pascolare 16, 00015 Monterotondo, Italy; walter.stefanoni@crea.gov.it (W.S.); francesco.latterini@crea.gov.it (F.L.); nadia.palmieri@crea.gov.it (N.P.); luigi.pari@crea.gov.it (L.P.)

2 Camelina Company Espana, Camino de la Carrera, 11, Fuente el Saz de Jarama, 28140 Madrid, Spain; jprieto@camelinacompany.es

3 Consiglio per la Ricerca in Agricoltura e l'Analisi dell'Economia Agraria (CREA), Centro di Ricerca Ingegneria e Trasformazioni Agroalimentari, Via Milano 43, 24047 Treviglio, Italy

* Correspondence: simone.bergonzoli@crea.gov.it

Abstract: The growing demand in food and non-food industries for camelina oil is driving the interest of farmers and contractors in investing in such feedstock. Nonetheless, the cost, performance and critical aspects related to the harvesting stage are still not properly investigated. In the present study, an ad-hoc test was performed in Spain in order to fulfill this gap. The results support the hypothesis to harvest camelina seeds with the same combine harvester used for cereal harvesting without further investment. Theoretical field capacity (TFC), effective field capacity (EFC), material capacity (MC), and field efficiency (FE) were 4.34 ha h^{-1}, 4.22 ha h^{-1}, 4.66 Mg h^{-1} FM, and 97.24%, respectively. The harvesting cost was estimated in 48.51 € ha^{-1}. Approximately, the seed loss of 0.057 ± 0.028 Mg ha^{-1} FM was due to the impact of the combine harvester header and dehiscence of pods, whilst 0.036 ± 0.006 Mg ha^{-1} FM of seeds were lost due to inefficiency of the threshing system of the combine harvester. Adjustment of the working speed of the combine and the rotation speed of the reel may help to reduce such loss.

Keywords: work productivity; harvesting costs; harvesting efficiency; wheat header; seed loss; header impact

Citation: Stefanoni, W.; Latterini, F.; Ruiz, J.P.; Bergonzoli, S.; Palmieri, N.; Pari, L. Assessing the Camelina (*Camelina sativa* (L.) Crantz) Seed Harvesting Using a Combine Harvester: A Case-Study on the Assessment of Work Performance and Seed Loss. *Sustainability* **2021**, *13*, 195. https://doi.org/10.3390/su13010195

Received: 30 October 2020
Accepted: 23 December 2020
Published: 28 December 2020

1. Introduction

The European Union is currently fostering the replacement of fossil-based products with bio-based surrogates [1,2]. Oil crops play a key role concerning this issue, thanks to their suitability to synthesize molecular structures which could be used to displace substantial amount of petroleum oil derived compounds [3,4]. Worldwide production of vegetable oil is given for 75% by few crops, such as soybean, oil palm, cottonseed, rapeseed and sunflower; while the remaining 25% is obtained from other minor oilseeds [1]. On the other hand, some of these minor oilseeds show particular features, which make them particularly suitable in the concept of bio-economy. In particular, camelina (*Camelina sativa* (L.) Crantz) belonging to *Brassicaceae* family [5] and originating from South-East Europe and South-West Asia [6], is a very promising oil crop for multiple reasons [7]. Camelina oil can indeed be used as edible oil rich in omega-3 fatty acids [8], and its oil and meal are also suitable sources of protein for both fish and ruminant diets [9–12]. Camelina oil has also multiple industrial applications, such as biodiesel and jet-fuel production, even if with some drawbacks related to cetane number, iodine value, oxidation stability and linolenic acid methyl ester content [13,14]. Furthermore, camelina oil can be used in the production of plasticizers, lubricants, polyols, resins, composites, coatings, elastomers, and adhesives [15]. Other interesting features of camelina are related to its cultivation. This species is indeed resistant to both drought and frost stress [16]. It has low nutritional requirements [17–19], with subsequent positive effects on the environment highlighted by life cycle assessment (LCA) studies [20,21], and can be grown on poor soils, also in a Mediterranean context [22], even if both seed yield and oil yield show substantial variability, i.e., 1.0–3.0 Mg ha^{-1} and 30–49% *w/w* respectively [23]. Finally, camelina, considering the presence of both winter and spring cultivar and the relatively brief life cycle, is suitable for double cropping with small grain cereals, soybean, and sunflower [24–29].

On the other hand, one of the main issues in camelina cultivation is the high costs of the supply chain [30]. Indeed, the higher percentage of costs for biodiesel production are related to the feedstock [31] and optimizing harvesting operation can lead to a substantial decrease of such costs [32].

According to this, costs of harvesting and logistic have to be evaluated, in order to make camelina cultivation fully sustainable and give support in the decision-making process to farmers and other stakeholders. Currently, mechanical harvesting of camelina is mainly carried out by using a combine harvester equipped with wheat header [33], only few experiences on cutting and swathing are reported [34]. However, seed loss can be very high, as a consequence of the tiny dimension of the seeds which are very small and light in weight [35,36] moreover, presence of weeds can further increase seed loss amount. Indeed, the entrance within the combine harvester of the green material of weeds, which generally shows higher moisture content than camelina, can reduce the efficiency of the threshing and cleaning system of the combine harvester, leading to higher seed loss [37]. Considering this, appropriate setting of the combine harvester and adjustment of working speed are fundamental to reduce seed loss [38]. However, combine harvester settings are not the only important aspects to take into account. In fact, camelina suffers seed loss for shattering as the ripeness is completed. Pods can easily open as consequence of external mechanical input, as the cutting bar of the combine harvester can provide. Hence, it is also important to finely regulate the rotary speed of the reel as well as the working speed of the machine in order to reduce such a phenomenon as much as possible. Some authors also suggest to consider the swathing method for harvesting in case of uneven ripeness [38].

Notwithstanding the centrality of this topic in the optic of a sustainable cultivation of camelina, few studies have focused on the evaluation of work performance, harvesting costs, and seeds loss.

The only comprehensive study reported in literature is Stefanoni et al. (2020), who reported a work productivity of 3.17 ha h^{-1}, with harvesting costs of 65.97 € ha^{-1} and seed loss of 7.82% *w/w* for a John Deere combine harvester (John Deere, Moline, IL, USA) [39]. In a previous work related to harvesting loss evaluation using a plot combine Sintim et al. (2016) found seed loss of 11.60% *w/w* [40], while Stolarski et al. (2019) reported harvesting cost per surface unit of 46.70 € ha^{-1} with a New Holland (New Holland, PA, USA) combine harvester [41].

Considering what is written above, there is still a need to investigate such a topic with specific field tests, in order to fill the knowledge gap that still exists. The aim of the present work is properly to provide the literature with significant information for both farmers and contractors; about work performance, costs and seed loss when collecting camelina seeds by combine harvester.

2. Materials and Methods

2.1. Experimental Field

Harvesting test was performed in the town of Astudillo, Palencia (Castilla y Leon, Spain) during the 27th week of 2020 (Figure 1). The experimental field (WGS84-UTM30T coordinates 390,896 E; 4,661,826 N) was flat and it measured 24.00 ha in surface and 893 m a.s.l in altitude.

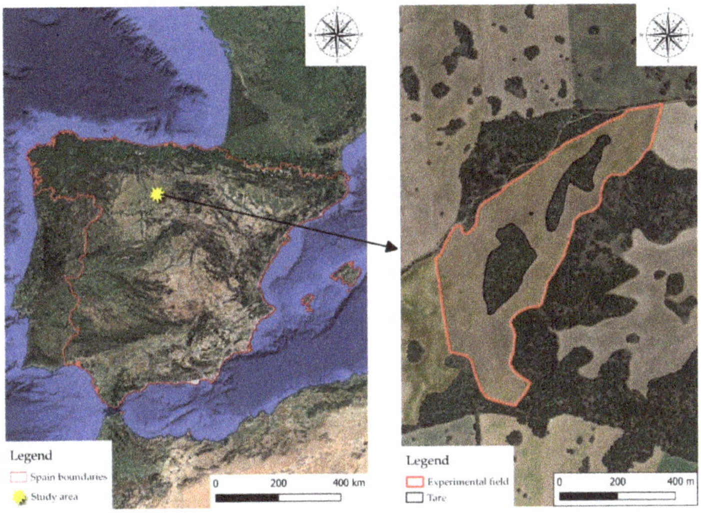

Figure 1. Experimental field location.

Camelina cultivation in the experimental field was carried out in conventional farming regime. Cultivar *Alba* (commercial variety provided by Camelina Company España) was sown in the first half of December 2019 with a seeding rate of 8 kg ha^{-1}. The previous crop was Barley. Fertilization was provided two times, with a rate of 250 kg ha^{-1} of NPK 8-15-15 in winter using a trailed fertilizer spreader and 250 kg ha^{-1} of liquid Nitrogen fertilizer (32%) in April by means of a mounted liquid fertilizer spreader. Chemical control of weeds was carried out before the nitrosulphate ammonium distribution by using a graminicide (Pilot, Quizalofop-p-ethyl 10%) to control the narrow-leaf weeds.

2.2. Pre-Harvest Test

Prior to the harvesting operation, 10 squared sample plots of 1 m^2 each were randomly established in order to assess the amount of the whole epigeous biomass (straw, siliques, and seeds). Camelina plants were cut at ground level with a shear, counted and then measured in both weight and height. Siliques and seeds were pulled and weighed separately. Consequently, siliques, seeds and a sample of straw from each plot were closed in sealed bags and transferred to the laboratory of Research Centre for Engineering and Agro-Food Processing (CREA-IT, Monterotondo, Rome, Italy) in order to perform further analysis. In particular, potential seed yield (PSY), dry weight (DW), 1000 seed-weight, bulk density and moisture content were evaluated. Dry weight and moisture content were estimated according to EN ISO 18134-2:2017 standard [42]. Seeds bulk density (kg m^{-3}) was calculated according to ISO 17828:201 [43] in 15 randomly selected samples.

2.3. Combine Harvester Model and Setting

The harvesting machinery was provided by the contractor. In particular the operation was carried out with a Claas Lexion 570 (Westfalia, Harsewinkel, Germany) combine harvester equipped with a conventional cleaning shoe and a 6.6 m wide cereal header. The machine had 273 kW diesel engine and the applied setting was as follow: rotor speed 800 rpm, cleaning fan speed 700 rpm, opening of the upper sieve 5/22 mm while lower sieve was closed. The combine harvester was moreover equipped with a straw chopper system, to thresh the straw and spread it on the ground.

2.4. Work Productivity

Harvesting productivity was tested in 6 sample plots randomly established in the study area. The area of each plot ranged between 420 to 950 m^2, and the evaluation of the working times was performed according to the methodology developed by Reith et al. (2017) [44]. The investigated parameters were: working speed (km h^{-1}), Theoretical Field Capacity (TFC, ha h^{-1}, calculated knowing the working speed and the width of the header), Effective Field Capacity (EFC, ha h^{-1}, calculated taking into account accessory times) and Material Capacity (MC, Mg h^{-1}, calculated knowing the EFC and the effective seed yield). The percentage ratio between EFC and TFC is named field efficiency (FE, %).

After harvesting operation, the collected material was unloaded onto a trailer and transported to the farm scale in order to be weighted.

2.5. Cost Analysis

Purchase and operating costs of the machinery were obtained interviewing the contractor, whilst the work productivity of the combine harvester was derived from the results of field tests and standard values for calculation were obtained from CRPA (Research Centre on Animal productions) methodology [45] as reported in Suardi et al. (2020) [46–48].

Hourly costs of harvesting machinery were calculated taking into account the market value of the combine harvester. The price of the combine harvester was discounted to 2019, using the lending rate of 3% provided by Banca d'Italia [49]. The parameter used for cost analysis are given in Table 1.

Table 1. Applied parameters for cost analysis.

	Parameter	Measure Unit	Value
Machine	Power	kW	240
Financial costs	Investment	€	362,615
	Service life	year	10
	Service life	H	3000
	Resale	%	19.00
	Resale	€	68,896.85
	Depreciation	€	293,718.15
	Annual usage	h $year^{-1}$	312
	Interest rate	%	3
Fixed costs	Ownership costs	€ $year^{-1}$	29,371.82
	Interests	€ $year^{-1}$	6472.67775
	Machine shelter	m^2	35.64
	Value of the shelter	€ m^{-2}	100
	Value of the shelter	€ $year^{-1}$	71.28
	Insurance	€ $year^{-1}$	906.5375
Variable costs	Repair factor	%	40
	Repairs and maintenance	€ h^{-1}	50.28
	Fuel cost	€ l^{-1}	0.57
	Fuel consumption	l h^{-1}	42.50
	Fuel cost	€ h^{-1}	24.23
	Lubricant cost	€ l^{-1}	3.03
	Lubricant consumption	l h^{-1}	0.38
	Lubricant cost	€ h^{-1}	1.14
	Worker salary	€ h^{-1}	11.5

2.6. Seed Loss Evaluation

Camelina seed loss was evaluated by counting the number of the seeds lying on the ground after the passage of the combine harvester. Specifically, two different areas behind the machine were selected as shown in Figure 2a: (A) in correspondence of the swath; (B) beside the swath but within the maximum cutting bar width. Ten squared sampling plots

10 cm × 10 cm (Figure 2b) were randomly selected within each region. Thus, in A, the seed loss was due to natural shattering (SS), impact of the header (ISL) and inefficiency of the cleaning shoe (CLS). On the other hand, in B, the seed loss was due to SS and ISL. Consequently, CLS was calculated as difference between the total seeds found in A and B regions. Since the loss due to CLS was concentrated in 1.6 m (the width of the swath), the difference in seed number between A and B was divided by 4.125 (the ratio between the cutting bar width and the swath width). By knowing the 1000-seed weight, the amount of seed loss was calculated in weight and referred to hectares.

Figure 2. On the left (**a**), identification of the areas A and B behind the combine harvester. On the right (**b**), example of a sample plot to detect seeds on the ground.

Furthermore, the effective seed loss (ESL) was also estimated by calculating the difference between the potential seed yield (PSY), measured in the pre-harvesting plot, and effective seed yield (ESY), measured by the farm scale after weighing the trailer.

The difference of the two methodologies was used to estimate the SS.

2.7. Statistical Analysis

The analysis of variance (ANOVA) was performed using the R 3.6.1 software to separate statistically different means among the groups ($p \leq 0.05$) [50].

3. Results

3.1. Pre-Harvest Test

Results of the pre-harvest tests are shown in Table 2. Before harvesting, 424 plants per m^2 were standing on the field and the mean plant height was 60 cm. Straw, siliques and seed moisture were 44.40%, 9.91%, and 6.45% respectively.

As reported in Figure 3, the largest aboveground portion was straw (69.62% *w/w*), then siliques and seeds (14.44% *w/w* and 15.94% *w/w* respectively). The harvest index (HI) was 0.223 and the potential seed yield was 1.17 Mg ha^{-1} FM.

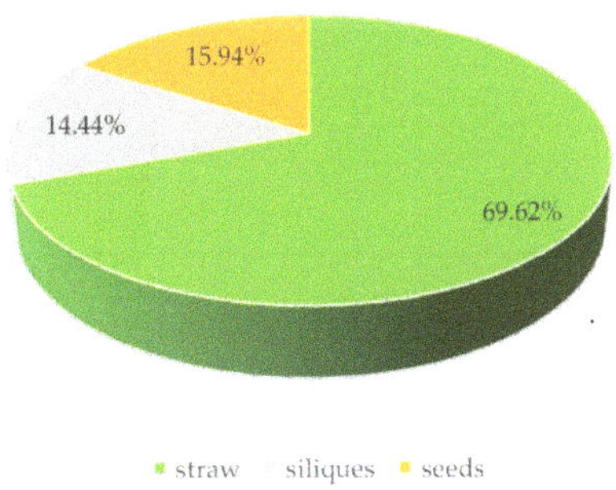

Figure 3. Percentage of straw, siliques and seeds of the aboveground biomass.

Table 2. Results of pre-harvest test reporting the mean quantity of available aboveground biomass, moisture content, and allocation among siliques, seeds and stalks. Weigh and bulk density of seeds is also reported.

Parameter	Measure Unit	Average	St.Dev.
Harvested surface	ha	24	-
Number of plants	N m^{-2}	424	176
Plant height	cm	60	8
Straw weight	Mg ha^{-1} FM	5.10	1.15
Straw moisture content	%	44.40	6.21
Siliques weight	Mg ha^{-1} FM	1.06	0.25
Siliques moisture content	%	9.91	0.49
Potential seed yield (PSY)	Mg ha^{-1} FM	1.17	0.18
Seed moisture content	%	6.45	0.40
1000-seed weight	g	1.04	0.07
Seed bulk density	kg m^{-3}	687.82	13.60

3.2. Work Productivity and Costs

Working performance of the combine harvester is reported in Table 3. The working speed was estimated being 6.57 km h^{-1}, while TFC and EFC were 4.34 ha h^{-1} and 4.22 ha h^{-1}, respectively. Considering the effective seed yield (1.10 Mg ha^{-1} FM), the MC and FE resulted in 4.66 Mg h^{-1} FM and 97.24%, respectively.

Table 3. Evaluation of the work performance of the combine harvester: theoretical and effective field capacity, field efficiency and material capacity.

Parameter	Measure Unit	Average	St.Dev.
Working speed	km h^{-1}	6.57	1.00
Theoretical Field Capacity (TFC)	ha h^{-1}	4.34	0.66
Effective Field Capacity (EFC)	ha h^{-1}	4.22	0.63
Field Efficiency (FE)	%	97.24	0.41
Material capacity (MC)	Mg h^{-1}	4.66	0.69

The analysis of working performance allowed to estimate the harvesting costs which were: 205.17 € h^{-1}, 48.51 € ha^{-1} and 43.92 € Mg^{-1} FM.

3.3. Seed Loss Evaluation

The seed loss calculated for each source of is reported in Table 4. TSL was 0.093 ± 0.033 Mg ha^{-1}, or 7.95 ± 0.28% of PSY. The majority of seed loss (4.87 ± 2.35% *w/w*) is linked to the impact of the header and the natural shattering. The latter further estimated in 1.97% *w/w* of the PSY as difference between TSL and ESL. On the other hand, CLS accounted for 3.08 ± 0.54% *w/w* of the TSL. The effective seed loss measured as the mere difference between the PSY and ESY, was 0.07 Mg ha^{-1}.

Table 4. Seed loss assessment according to the two methodologies. Common letters within columns denote the absence of significant difference ($p < 0.05$).

Parameter	Average	
	Mg ha^{-1} FM	%
Area A (CSL)	0.036 ± 0.006 b	3.08 ± 0.54
Area B (SS+ISL)	0.057 ± 0.028 a	4.87 ± 2.35
Total seed loss (TSL)	0.093 ± 0.033	7.95 ± 0.28
Potential Seed Yield (PSY)	1.17 ± 0.18	
Effective Seed Yield (ESY)	1.10 *	
Effective seed loss (ESL)	0.07 *	5.98 *

Note: (*) this value was not replicated since all grains were collected within one trailer and weighted only once at the end of the harvesting.

4. Discussions

4.1. Aboveground Biomass Yield

The potential seed yield assessed in the pre-harvest test of 1.17 ± 0.18 Mg ha^{-1} FM (seed moisture content of 6.45 ± 0.40%) is in line with the findings reported in Mauri et al. (2019) and Stefanoni et al. (2020) for similar experiments conducted in Spain [39,51] as well as in USA as reported by Schillinger et al. 2019 [52]. Higher values of seed yield are reported by Royo-Esnal et al. (2018) in Eastern Spain with seed yield ranging from 0.92 to 2.31 Mg ha^{-1} FM after a comparison of different sowing rates: 8 kg ha^{-1} and 11 kg ha^{-1} [27]. However, the authors did not find a significant effect of the sowing rate upon potential seed yield, nor with the weed coverage. Similarly, Zanetti et al. (2020) reported a negligible effect of the plant density on seed yield, whilst later sowing could improve oil content [22]. In the present study, instead, fertilizer and chemicals were used to both providing nutrients and controlling the weeds. In similar studies, where camelina was grown under conventional farming, the potential seed yield doubled in comparison with not fertilized fields (namely, 0.93 Mg ha^{-1} FM and 1.81 Mg ha^{-1} FM) [53]. Comparing with other herbaceous oilseeds, for instance, camelina performs slightly lower than castor (*Ricinus communis* L. up to 4.4 Mg ha^{-1}), canola (*Brassica napus* L. 2.19 Mg ha^{-1} FM), sunflower (*Heliantus annuus* L. 1.97 Mg ha^{-1} FM) [33], although it is suitable for cropping in marginal land [54]. After harvesting, seeds usually face some storing which could be also long in time before being processed. This condition can lead to low quality product, or even loss of the entire product if moisture is too high. In the present study, seed moisture was found as low as 6.45 ± 0.4% which is far below the threshold of 8% as reported by [55]. 1000 seed-weight was also recorded and it averaged to 1.04 ± 0.07 g in fresh weight which is consistent with the value found by other authors [39]. If compared with other Brassicaceae family and seed weight, it is rather low. In fact, Kuai et al. (2015) reported 3.3 and 3.5 g in rapeseed (*Brassica napus* L.) [56], Zhu et al. (2016) reported values ranging from 6.0 to 9.5 g per thousand seeds in crambe (*Crambe abyssinica*) [57].

Despite the seeds that find application on both food and non-food sectors, straw and siliques from camelina (5.10 ± 1.15 Mg ha^{-1} FM and 1.06 ± 0.25 Mg ha^{-1} FM respectively) can also be attractive for energy industry. In fact, they both are valid feedstocks for bioenergy production via pyrolysis due to the low nitrogen content (0.4–0.5%) and the low char production (approximately 25.5%) [58]. However, the chemical-physical properties of camelina residual biomass can vary according to the growth conditions. For instance, camelina grown in the Central Italy exhibits high cellulose and hemicellulose content in comparison with camelina grown in the Northern Italy while the ash content is not affected by such factor [59]. This implies that different scenarios are opened for the exploitation of camelina residual biomass in a sustainable green chemistry approach. Moreover, the development of a proper value chain of the residual biomass may contribute to the reduction of greenhouse gas emissions that occur during the degradation of the organic matter in the soil as reported in other oil crops [60].

4.2. Work Productivity and Costs

In the present study, a conventional combine harvester equipped with a cereal header was used. The literature still lacks the knowledge on such kind of strategy for harvesting camelina seeds, therefore a comparison is possible relying on the findings reported in Stefanoni et al. (2020) [39]. Here, the working speed of the machine was 30.1% higher. This caused the increase of TFC, EFC, FE, and MC by 29.29%, 33.12%, 3.54%, and 54.82% respectively. Interestingly, the cutting bars of the combine harvesters measured 6.6 and 6.7 m wide. Such a negligible deviation leads to the conclusion that the difference in the performance are exclusively related to the different working speeds.

Interestingly, comparing the performance of the combine harvester found in the present study with those reported in similar studies but conducted on wheat grains harvesting, here again the findings are higher. Normally, TFC ranges from 2.61 ha h^{-1} to 3.72 ha h^{-1}, while the EFC is 1.92–2.28 ha h^{-1} and the FE is as high as 83% [46–48]. However, it is important to underline that such high working performance is related to the dimensions and to the particular shape of the experimental field, which allowed to minimize the turnings, thus decreasing the accessory time and increasing the EFC and FE.

The harvesting cost was assessed in 48.51 € ha^{-1} and 43.92 € Mg^{-1} FM which are consistent with the cost shown by Stolarski et al. (2019) [41], but much lower than that calculated in a similar harvesting trial performed in Spain on camelina crop (65.97 € ha^{-1} and 69.42 € Mg^{-1} FM) [39]. Other trials performed on wheat and corn grain harvesting with combine harvester showed harvesting costs being 77.98 and 129.51 € ha^{-1}, respectively [46,61].

4.3. Seed Loss Evaluation

The evaluation of the seeds loss during harvesting stage is an important parameter to take into account since it contributes to reduce the revenue of farmers and contractors therefore, the loss of seeds should be as low as possible. Generally, the amount of seed lost is calculated as the difference between the potential seed yield (1.17 $\pm$ 0.18 Mg ha^{-1} FM) and the effective seed yield (1.10 Mg ha^{-1} FM) which, in this specific trial, was 0.07 Mg ha^{-1} FM (5.98% *w/w*). Higher values were found by Stefanoni et al. (2020) and Sintim et al. (2016) which found 7.82 and 11.70% *w/w*, respectively [39,40]. In other herbaceous oil crops, seed loss ranges from 1% as in sunflower [62,63] or in canola [64,65], and 3% as in safflower [33], or even higher as in castor bean harvesting [33]. However, such information only provides evidence regarding the total amount of seed loss, but it fails in pointing out what is responsible for that loss. A combine harvester is a complicated machine which can generate different sources of loss particularly if seeds are small and light in weight (only 1.04 $\pm$ 0.07 g FM per 1000 seeds). The main sources of loss are the impact of the header and the inefficiency of the cleaning shoe. If some actions have to be taken against the seed loss in camelina harvesting, their respective contribution to the TSL must be investigated. According to Table 4, the inefficiency of the cleaning shoe of the combine harvester (CSL) triggered the loss of 0.036 $\pm$ 0.006 Mg ha^{-1} FM (3.08 $\pm$ 0.54% *w/w*) of the seeds, while 0.057 $\pm$ 0.028 Mg ha^{-1} FM (4.87 $\pm$ 2.35% *w/w*) of seeds were lost due SS and ISL. Interestingly, TSL and ESL differed for 0.023 Mg ha^{-1} FM (1.97% *w/w* of the PSY) which can be partially explained as the loss due to SS (natural pod shattering) which occurs spontaneously in camelina as it ripens. In fact, late harvesting can lead to a loss of seeds due to SS as high as 25% *w/w* in some cultivars [66]. Moreover, pod shattering can be triggered by a minimum external input in completely ripened pods. Therefore, the mechanical disturbance provided by the combine harvester can contribute significantly to increase such phenomenon, particularly as working and rotation speed of the reel (the latter value was not measured in the present study).

5. Conclusions

Camelina is gathering more and more attention throughout the Europe since its multipurpose oil as well as the aboveground suitability for bioenergy purposes. However, the related value chain is still not well developed, partially because the crucial phase of the harvest has not been comprehensively investigated so far. Our findings support the hypothesis that a combine harvester equipped with wheat header is suitable for camelina seed harvesting, which is particular convenient for farmers and contractor who use camelina as rotation crop in winter cereals since the same machine is valid for both crops. Furthermore, the cost and the performance are similar. Little concern may arise regarding the seed loss which are mainly linked to impact of the header of the combine harvester, and the inability of the cleaning shoe to efficiently discriminate the seeds from the other portions of the biomass. This latter problem can be partially addressed by simply reducing the speed of the machine. Instead, natural pod shattering contributes marginally to the loss of seeds.

Author Contributions: Conceptualization, W.S., F.L., S.B., and L.P.; methodology, W.S., F.L., S.B., and J.P.R.; data curation, W.S., F.L., S.B., N.P., and J.P.R.; writing—original draft preparation, W.S., F.L., S.B., and N.P.; writing—review and editing, W.S., F.L., S.B., L.P., and J.P.R.; supervision, L.P.; funding acquisition, L.P. All authors have read and agreed to the published version of the manuscript.

Funding: This research was funded by PRIMA foundation, project 4CE-MED, grant Number 1911 and by Horizon 2020 project Panacea, grant Number 773501.

Institutional Review Board Statement: Not applicable.

Informed Consent Statement: Not applicable.

Data Availability Statement: The data presented in this study are available on request from the corresponding author. The data are not publicly available due to restrictions e.g., privacy.

Acknowledgments: The authors wish to thank Camelina Company (Camino de la Carrera, 11-11, 28140 Fuente el Saz, Madrid, Spain) for the organization of the tests and the support during the trials and Sandu Lazar for the help in performing the field and laboratory tests.

Conflicts of Interest: The authors declare no conflict of interest.

References

1. Righini, D.; Zanetti, F.; Monti, A. The bio-based economy can serve as the springboard for camelina and crambe to quit the limbo. *OCL* **2016**, *23*, D504. [CrossRef]
2. Krzyżaniak, M.; Stolarski, M.J.; Graban, Ł.; Lajszner, W.; Kuriata, T. Camelina and Crambe Oil Crops for Bioeconomy—Straw Utilisation for Energy. *Energies* **2020**, *13*, 1503. [CrossRef]
3. de Espinosa, L.M.; Meier, M.A.R. Plant oils: The perfect renewable resource for polymer science? *Eur. Polym. J.* **2011**, *47*, 837–852. [CrossRef]
4. Carlsson, A.S. Plant oils as feedstock alternatives to petroleum—A short survey of potential oil crop platforms. *Biochimie* **2009**, *91*, 665–670. [CrossRef]
5. Natelson, R.H.; Wang, W.C.; Roberts, W.L.; Zering, K.D. Technoeconomic analysis of jet fuel production from hydrolysis, decarboxylation, and reforming of camelina oil. *Biomass Bioenergy* **2015**, *75*, 23–34. [CrossRef]
6. Larsson, M. Cultivation and processing of Linum usitatissimum and Camelina sativa in southern Scandinavia during the Roman Iron Age. *Veg. Hist. Archaeobot.* **2013**, *22*, 509–520. [CrossRef]
7. Krzyżaniak, M.; Stolarski, M.J.; Tworkowski, J.; Puttick, D.; Eynck, C.; Załuski, D.; Kwiatkowski, J. Yield and seed composition of 10 spring camelina genotypes cultivated in the temperate climate of Central Europe. *Ind. Crops Prod.* **2019**, *138*, 111443. [CrossRef]
8. Eidhin, D.N.; Burke, J.; O'Beirne, D. Oxidative stability of ω3-rich camelina oil and camelina oil-based spread compared with plant and fish oils and sunflower spread. *J. Food Sci.* **2003**, *68*, 345–353. [CrossRef]
9. Halmemies-Beauchet-Filleau, A.; Shingfield, K.J.; Simpura, I.; Kokkonen, T.; Jaakkola, S.; Toivonen, V.; Vanhatalo, A. Effect of incremental amounts of camelina oil on milk fatty acid composition in lactating cows fed diets based on a mixture of grass and red clover silage and concentrates containing camelina expeller. *J. Dairy Sci.* **2017**, *100*, 305–324. [CrossRef]
10. Hixson, S.M.; Parrish, C.C.; Wells, J.S.; Winkowski, E.M.; Anderson, D.M.; Bullerwell, C.N. Inclusion of camelina meal as a protein source in diets for farmed salmonids. *Aquac. Nutr.* **2016**, *22*, 615–630. [CrossRef]
11. Lawrence, R.D.; Anderson, J.L. Ruminal degradation and intestinal digestibility of camelina meal and carinata meal compared with other protein sources. *Prof. Anim. Sci.* **2018**, *34*, 10–18. [CrossRef]
12. Betancor, M.B.; MacEwan, A.; Sprague, M.; Gong, X.; Montero, D.; Han, L.; Napier, J.A.; Norambuena, F.; Izquierdo, M.; Tocher, D.R. Oil from transgenic Camelina sativa as a source of EPA and DHA in feed for European sea bass (*Dicentrarchus labrax* L.). *Aquaculture* **2021**, *530*, 735759. [CrossRef]
13. Ciubota-Rosie, C.; Ruiz, J.R.; Ramos, M.J.; Pérez, Á. Biodiesel from Camelina sativa: A comprehensive characterisation. *Fuel* **2013**, *105*, 572–577. [CrossRef]
14. Corporan, E.; Edwards, T.; Shafer, L.; DeWitt, M.J.; Klingshirn, C.; Zabarnick, S.; West, Z.; Striebich, R.; Graham, J.; Klein, J. Chemical, thermal stability, seal swell, and emissions studies of alternative jet fuels. *Energy Fuels* **2011**, *25*, 955–966. [CrossRef]

15. Kim, N.; Li, Y.; Sun, X.S. Epoxidation of Camelina sativa oil and peel adhesion properties. *Ind. Crops Prod.* **2015**, *64*, 1–8. [CrossRef]
16. Allen, B.L.; Vigil, M.F.; Jabro, J.D. Camelina growing degree hour and base temperature requirements. *Agron. J.* **2014**, *106*, 940–944. [CrossRef]
17. Berti, M.; Wilckens, R.; Fischer, S.; Solis, A.; Johnson, B. Seeding date influence on camelina seed yield, yield components, and oil content in Chile. *Ind. Crops Prod.* **2011**, *34*, 1358–1365. [CrossRef]
18. Johnson, J.M.F.; Gesch, R.W. Calendula and camelina response to nitrogen fertility. *Ind. Crops Prod.* **2013**, *43*, 684–691. [CrossRef]
19. Román-Figueroa, C.; Padilla, R.; Uribe, J.; Paneque, M. Land Suitability Assessment for Camelina (*Camelina sativa* L.) Development in Chile. *Sustainability* **2017**, *9*, 154. [CrossRef]
20. Krzyżaniak, M.; Stolarski, M.J. Life cycle assessment of camelina and crambe production for biorefinery and energy purposes. *J. Clean. Prod.* **2019**, *237*, 117755. [CrossRef]
21. Brentrup, F.; Küsters, J.; Lammel, J.; Kuhlmann, H. Methods to estimate on-field nitrogen emissions from crop production as an input to LCA studies in the agricultural sector. *Int. J. Life Cycle Assess.* **2000**, *5*, 349. [CrossRef]
22. Zanetti, F.; Gesch, R.W.; Walia, M.K.; Johnson, J.M.F.; Monti, A. Winter camelina root characteristics and yield performance under contrasting environmental conditions. *Field Crops Res.* **2020**, *252*, 107794. [CrossRef]
23. Berti, M.; Gesch, R.; Eynck, C.; Anderson, J.; Cermak, S. Camelina uses, genetics, genomics, production, and management. *Ind. Crops Prod.* **2016**, *94*, 690–710. [CrossRef]
24. Chen, C.; Bekkerman, A.; Afshar, R.K.; Neill, K. Intensification of dryland cropping systems for bio-feedstock production: Evaluation of agronomic and economic benefits of Camelina sativa. *Ind. Crops Prod.* **2015**, *71*, 114–121. [CrossRef]
25. Gesch, R.W.; Archer, D.W. Double-cropping with winter camelina in the northern Corn Belt to produce fuel and food. *Ind. Crops Prod.* **2013**, *44*, 718–725. [CrossRef]
26. Berti, M.; Samarappuli, D.; Johnson, B.L.; Gesch, R.W. Integrating winter camelina into maize and soybean cropping systems. *Ind. Crops Prod.* **2017**, *107*, 595–601. [CrossRef]
27. Royo-Esnal, A.; Valencia-Gredilla, F. Camelina as a Rotation Crop for Weed Control in Organic Farming in a Semiarid Mediterranean Climate. *Agriculture* **2018**, *8*, 156. [CrossRef]
28. Peterson, A.T.; Berti, M.T.; Samarappuli, D. Intersowing cover crops into standing soybean in the US upper midwest. *Agronomy* **2019**, *9*, 264. [CrossRef]
29. Zanetti, F.; Christou, M.; Alexopoulou, E.; Berti, M.T.; Vecchi, A.; Borghesi, A.; Monti, A. Innovative double cropping systems including camelina [*Camelina sativa* (L.) crantz] a valuable oilseed crop for bio-based applications. In Proceedings of the European Biomass Conference and Exhibition Proceedings, Lisbon, Portugal, 27–30 May 2019; pp. 127–130.
30. Christopher, L.P.; Kumar, H.; Zambare, V.P. Enzymatic biodiesel: Challenges and opportunities. *Appl. Energy* **2014**, *119*, 497–520. [CrossRef]
31. Bušić, A.; Kundas, S.; Morzak, G.; Belskaya, H.; Mardetko, N.; Šantek, M.I.; Komes, D.; Novak, S.; Šantek, B. Recent trends in biodiesel and biogas production. *Food Technol. Biotechnol.* **2018**, *56*, 152–173. [CrossRef]
32. Latterini, F.; Stefanoni, W.; Suardi, A.; Alfano, V.; Bergonzoli, S.; Palmieri, N.; Pari, L. A GIS Approach to Locate a Small Size Biomass Plant Powered by Olive Pruning and to Estimate Supply Chain Costs. *Energies* **2020**, *13*, 3385. [CrossRef]
33. Pari, L.; Latterini, F.; Stefanoni, W. Herbaceous Oil Crops, a Review on Mechanical Harvesting State of the Art. *Agriculture* **2020**, *10*, 309. [CrossRef]
34. McVay, K.A.; Lamb, P.F. Camelina production in Montana. *Bull. MT200701AG Mont. State Univ.* **2008**.
35. Zubr, J. Qualitative variation of Camelina sativa seed from different locations. *Ind. Crops Prod.* **2003**, *17*, 161–169. [CrossRef]
36. Vollmann, J.; Moritz, T.; Kargl, C.; Baumgartner, S.; Wagentristl, H. Agronomic evaluation of camelina genotypes selected for seed quality characteristics. *Ind. Crops Prod.* **2007**, *26*, 270–277. [CrossRef]
37. Leclère, M.; Jeuffroy, M.-H.; Butier, A.; Chatain, C.; Loyce, C. Controlling weeds in camelina with innovative herbicide-free crop management routes across various environments. *Ind. Crops Prod.* **2019**, *140*, 111605. [CrossRef]
38. Obour, K.A.; Sintim, H.J.; Obeng, E.; Jeliazkov, V.D. Oilseed Camelina (*Camelina sativa* L. Crantz): Production Systems, Prospects and Challenges in the USA Great Plains. *Adv. Plants Agric. Res.* **2015**, *2*, 00043. [CrossRef]
39. Stefanoni, W.; Latterini, F.; Ruiz, J.P.; Bergonzoli, S.; Attolico, C.; Pari, L. Mechanical Harvesting of Camelina: Work Productivity, Costs and Seed Loss Evaluation. *Energies* **2020**, *13*, 5329. [CrossRef]
40. Sintim, H.Y.; Zheljazkov, V.D.; Obour, A.K.; Garcia y Garcia, A. Managing harvest time to control pod shattering in oilseed camelina. *Agron. J.* **2016**, *108*, 656–661. [CrossRef]
41. Stolarski, M.J.; Krzyżaniak, M.; Tworkowski, J.; Załuski, D.; Kwiatkowski, J.; Szczukowski, S. Camelina and crambe production—Energy efficiency indices depending on nitrogen fertilizer application. *Ind. Crops Prod.* **2019**, *137*, 386–395. [CrossRef]
42. NSAI ISO 18134-2:2017 Solid Biofuels—Determination of Moisture Content—Oven Dry Method—Part 2: Total Moisture—Simplified Method. Available online: https://www.iso.org/standard/71536.html (accessed on 6 July 2020).
43. ISO 17828:2015 ISO 17828:2015 Solid Biofuels—Determination of Bulk Density. Available online: https://www.iso.org/obp/ui/#iso:std:iso:17828:ed-1:v1:en (accessed on 23 July 2020).
44. Reith, S.; Frisch, J.; Winkler, B. Revision of the working time classification to optimize work processes in modern agriculture. *Chem. Eng. Trans.* **2017**, *58*, 121–126. [CrossRef]
45. Assirelli, A.; Pignedoli, S. Costo di esercizio delle macchine agricole. *Cent. Ric. Prod. Anim.* **2005**, *5*, 1–10.

46. Suardi, A.; Stefanoni, W.; Bergonzoli, S.; Latterini, F.; Jonsson, N.; Pari, L. Comparison between Two Strategies for the Collection of Wheat Residue after Mechanical Harvesting: Performance and Cost Analysis. *Sustainability* **2020**, *12*, 4936. [CrossRef]
47. Suardi, A.; Stefanoni, W.; Alfano, V.; Bergonzoli, S.; Pari, L. Equipping a Combine Harvester with Turbine Technology Increases the Recovery of Residual Biomass from Cereal Crops via the Collection of Chaff. *Energies* **2020**, *13*, 1572. [CrossRef]
48. Suardi, A.; Saia, S.; Stefanoni, W.; Gunnarsson, C.; Sundberg, M.; Pari, L. Admixing Chaff with Straw Increased the Residues Collected without Compromising Machinery Efficiencies. *Energies* **2020**, *13*, 1766. [CrossRef]
49. Banca d'Italia Banca d'Italia Lending Rate. Available online: https://www.bancaditalia.it/ (accessed on 11 July 2020).
50. R Development Core Team. R: A Language and Environment for Statistical Computing. R Foundation for Statistical Computing: Vienna, Austria, 2010. Available online: http://www.r-project.org/ (accessed on 6 August 2020).
51. Mauri, P.V.; Mostaza, D.; Plaza, A.; Ruiz-Fernandez, J.; Prieto, J.; Capuano, A. Variability of camelina production in the center of Spain in two years of cultivation, a new profitable and alternative crop. In Proceedings of the 27th European Biomass Conference and Exhibition, Lisbon, Portugal, 27–30 May 2019; pp. 196–200.
52. Schillinger, W.F. Camelina: Long-term cropping systems research in a dry Mediterranean climate. *Field Crop. Res.* **2019**, *235*, 87–94. [CrossRef]
53. Imbrea, F.; Jurcoane, S.; Hălmăjan, H.V.; Duda, M.; Botoş, L. Camelina sativa: A new source of vegetal oils. *Rom. Biotechnol. Lett.* **2011**, *16*, 6263–6270.
54. Lohaus, R.H.; Neupane, D.; Mengistu, M.A.; Solomon, J.K.Q.; Cushman, J.C. Five-Year Field Trial of Eight Camelina sativa Cultivars for Biomass to be Used in Biofuel under Irrigated Conditions in a Semi-Arid Climate. *Agronomy* **2020**, *10*, 562. [CrossRef]
55. Grady, K.; Nleya, T. Camelina Production. *Ext. Extra South Dakota State Univ.* **2010**, *8167*, 1–3.
56. Kuai, J.; Sun, Y.; Zuo, Q.; Huang, H.; Liao, Q.; Wu, C.; Lu, J.; Wu, J.; Zhou, G. The yield of mechanically harvested rapeseed (*Brassica napus* L.) can be increased by optimum plant density and row spacing. *Sci. Rep.* **2015**, *5*, 1–14. [CrossRef]
57. Zhu, L.H.; Krens, F.; Smith, M.A.; Li, X.; Qi, W.; Van Loo, E.N.; Iven, T.; Feussner, I.; Nazarenus, T.J.; Huai, D.; et al. Dedicated Industrial Oilseed Crops as Metabolic Engineering Platforms for Sustainable Industrial Feedstock Production. *Sci. Rep.* **2016**, *6*, 1–11. [CrossRef] [PubMed]
58. Gómez-Monedero, B.; Bimbela, F.; Arauzo, J.; Faria, J.; Ruiz, M.P. Pyrolysis of Red Eucalyptus, Camelina Straw, and Wheat Straw in an Ablative Reactor. *Energy Fuels* **2015**, *29*, 1766–1775. [CrossRef]
59. Pagnotta, E.; Ugolini, L.; Matteo, R.; Lazzeri, L.; Foschi, L.; Angelini, L.G.; Tavarini, S. Exploring the Camelina sativa value chain: A new opportunity for bio-based products and overall crop sustainability. *Riv. Ital. Delle Sostanze Grasse* **2019**, *96*, 259–268.
60. Pari, L.; Suardi, A.; Stefanoni, W.; Latterini, F.; Palmieri, N. Environmental and Economic Assessment of Castor Oil Supply Chain: A Case Study. *Sustainability* **2020**, *12*, 6339. [CrossRef]
61. Bergonzoli, S.; Suardi, A.; Rezaie, N.; Alfano, V.; Pari, L. An innovative system for Maize Cob and wheat chaff harvesting: Simultaneous grain and residues collection. *Energies* **2020**, *13*, 1265. [CrossRef]
62. Startsev, A.S.; Makarov, S.A.; Nesterov, E.S.; Kazakov, Y.F.; Terentyev, A.G. Comparative evaluation of the operation of a combine harvester with an additional sieve with adjustable holes for sunflower harvesting. *IOP Conf. Ser. Earth Environ. Sci.* **2020**, *433*. [CrossRef]
63. Shaforostov, V.D.; Makarov, S.S. The header for a breeding plot combine for sunflower harvesting. *Acta Technol. Agric.* **2019**, *22*, 60–63. [CrossRef]
64. Asoodar, A.M.; Izadinia, Y.; Desbiolles, J. Benefits of harvester front extension in reducing canola harvest losses. *Int. Agric. Eng. J.* **2012**, *21*, 32–37.
65. Pari, L.; Assirelli, A.; Suardi, A.; Civitarese, V.; Del Giudice, A.; Costa, C.; Santangelo, E. The harvest of oilseed rape (*Brassica napus* L.): The effective yield losses at on-farm scale in the Italian area. *Biomass Bioenergy* **2012**, *46*, 453–458. [CrossRef]
66. Hobson, R.N.; Bruce, D.M. Seed loss when cutting a standing crop of oilseed rape with two types of combine harvester header. *Biosyst. Eng.* **2002**, *81*, 281–286. [CrossRef]

sustainability

MDPI

Article

Inulin Content in Chipped and Whole Roots of Cardoon after Six Months Storage under Natural Conditions

Luigi Pari [1], Vincenzo Alfano [1,*], Walter Stefanoni [1], Francesco Latterini [1], Federico Liuzzi [2], Isabella De Bari [2], Vito Valerio [2] and Anna Ciancolini [3]

1 Consiglio per la Ricerca in Agricoltura e l'Analisi dell'Economia Agraria (CREA)—Centro di Ricerca Ingegneria e Trasformazioni Agroalimentari, Via della Pascolare, 16, 00015 Monterotondo, Italy; luigi.pari@crea.gov.it (L.P.); walter.stefanoni@crea.gov.it (W.S.); francesco.latterini@crea.gov.it (F.L.)

2 ENEA Centro Ricerche Trisaia, 75026 Rotondella, Italy; federico.liuzzi@enea.it (F.L.); isabella.debari@enea.it (I.D.B.); vito.valerio@enea.it (V.V.)

3 NOVAMONT SpA, 28100 Novara, Italy; anna.ciancolini@novamont.com

* Correspondence: vincenzo.alfano@crea.gov.it; Tel.: +39-06-9067-5315

Abstract: Industries currently rely on chicory and Jerusalem artichoke for inulin extraction but also cardoon is proved to synthetize and store high quantity of inulin in roots as well. Cardoon is a multipurpose crop, well adapted to marginal lands, whose main residues at the end of cropping cycle consist of roots. However, cardoon roots are a suitable source of inulin, that is of high interest for new generation biodegradable bioplastics production. On the other hand, a sustainable supply chain for inulin production from cardoon roots has not been developed yet. In particular, in the inulin supply chain the most critical part is storage, which can negatively affect both cost and inulin quantity. In the present study the effect on inulin content in cardoon roots stored as dried chipped roots (CRt) and dried whole roots (WRt) was investigated in a 6-month storage trial. Our findings suggest that chipping before storage did not affect the inulin content during the storage. Furthermore, it reduced the time needed for drying by 33.3% and increased the bulk density by 154.9% with the consequent reduction of direct cost for drying, transportation and storage.

Keywords: Cynara roots; biorefinery; marginal lands; multipurpose crop; fermentable sugars; agricultural residues exploitation

Citation: Pari, L.; Alfano, V.; Stefanoni, W.; Latterini, F.; Liuzzi, F.; De Bari, I.; Valerio, V.; Ciancolini, A. Inulin Content in Chipped and Whole Roots of Cardoon after Six Months Storage under Natural Conditions. *Sustainability* **2021**, *13*, 3902. https://doi.org/10.3390/su13073902

Academic Editor: Anastasios Michailidis

Received: 12 March 2021
Accepted: 29 March 2021
Published: 1 April 2021

1. Introduction

Cardoon (*Cynara cardunculus* L.) is one of the most promising feedstocks for biorefinery in the Mediterranean areas since it is a multipurpose perennial crop well adapted to drought environments and low productive marginal lands [1–4]. The cultivation of this species mainly focuses on the exploitation of the aerial biomass as seeds, leaves and stalks [5–7]. The vegetable oil extracted from seeds is rich in monounsaturated fatty acids useful to produce important intermediates such as azelaic acid or pelargonic acid, that are highly demanded by synthetic fertilizer industries as well as cosmetic industries worldwide [8,9]. On the other hand, leaves and stalks represent an important source of lignocellulosic biomass potentially suitable for the production of intermediate compounds, like bioethanol and Bio-butanediol, which are widely used for producing bioplastics [10–12].

However, the potential of cardoon as a multipurpose crop has not been fully exploited yet, in particular regarding the presence in the roots of inulin suitable for nutraceutical, pharmaceutical and other biorefining applications [13–15].

Inulin is a linear fructan, i.e., a polymer of fructose units linked by β (2 → 1) glycosidic bonds with a variable degree of polymerization (DP), between 3 and 60, and usually a glucose molecule at the end [16]. Inulin can be used in food and pharmaceutical industry for several purposes, such as prebiotics to stimulate the growth of probiotic gut bacteria, for nutritional purposes as low caloric soluble dietary fiber and also as a mediate sugar and lipid metabolism in diabetic and hypercholesterolemia [17]. In medicine, inulin is also used as a diagnostic agent for the determination of kidney function [18]. In the biorefinery industry, inulin and inulin-rich biomass are gaining interest for the production of fructose by enzymatic hydrolysis of inulin, as alternative way to the current approaches based on acid hydrolysis of sucrose [19,20]. In the biorefinery, moreover, the availability of fermentable sugars is crucial to produce ethanol, and inulin is a good feedstock for bioethanol production by fermentation after hydrolysis [21,22].

Inulin is a reserve carbohydrate accumulated mainly in the roots and tubers of many plants belonging to the Asteraceae family, like Cardoon [16].

Among the Asteraceae family plants, Chicory (*Cichorium intybus* L.) and Jerusalem artichoke (*Helianthus tuberosus* L.) are currently the major industrial sources of inulin [23]. In a comparative study aimed at evaluating different types of inulin, extracted from Cardoon roots, Jerusalem artichoke tubers and Chicory roots, the inulin amount resulted respectively in 115, 390 and 550 g kg^{-1} of d.m. [24].

In the perspective of new generation biorefineries and the circular bioeconomy concept, the recovery of inulin from cardoon roots at the end of crop cycle, in addition to various high added-value raw materials from seeds and stalks, seems to be an interesting opportunity. However, the full development of an effective value chain for the biochemical industries, implies well-organized logistics. Storage phase, in particular, has a high impact on the quality of the raw material and on the overall costs of the value chain [25].

Effects of storage conditions on inulin content have been investigated in different inulin-containing crops. In the case of the storage of Jerusalem artichoke tubers, inulin composition remained stable under frozen storage (−18 °C) during 3 months of study, while a significant degradation of inulin to sucrose and fructo-oligosaccharides was observed after 4 weeks when the storage was performed at 4 °C [26]. In another study, inulin hydrolase activity in the tubers of Jerusalem artichoke peaked at the 15th day of storage at ambient conditions: inulin underwent depolymerization causing a decrease in inulin content and an increase in soluble sugars [27]. In a 28 day storage trial a decrease of 70% and 96% was reported for artichoke heads after storage at 4 and 18 °C, respectively. Similarly, in storage of sliced artichoke heads at 4 °C, about 60% decrease of inulin content was observed during the first 11 days of storage [28].

In addition to the temperature, moisture plays a key role to start the enzymatic hydrolysis of the inulin. For example, it has been observed that the enzymatic hydrolysis of the inulin during storage of chicory roots depends on the moisture content, while the generated sugars favored the loss of material as a result of cell respiration and microbial activity [29]. In fact, microorganisms require minimum thresholds of moisture to maintain and optimize metabolism, that is the breakdown and the consumption of the sugar-based components of dry matter. Moisture below 10% is generally considered to be low enough to prevent microbial degradation and allow for safe long-term storage of biomass [30]. In this framework, the thermal drying technologies have gained interest as effective tools for extending the length of storage as well as reducing the handling cost and ease the transportation that affect the value chain at the industrial level [31,32].

On the other hand, a suitable storage system for a biorefinery has not only to focus on the capacity of keeping a high content of a given product, but there is also the need of finding a suitable solution regarding the economic sustainability, with particular reference to transport costs which can have a substantial impact on the overall value chain [33,34]. Under this point of view biomass chipping is an interesting approach. Indeed, the higher bulk density achieved by the chipped material improves the logistics by reducing the space needed during transport as well as in the storage area [35]. On the other hand, chipping increases the exposed surface area and reduces the air permeability [36] with an expectable opposite effect on drying efficiency as well as on the maintenance of dry matter and inulin content.

In order to set up a suitable value chain for inulin production from cardoon roots biomass, there is the need of investigating a storage system which combines effective long-term maintenance of inulin and economic sustainability. Considering the current lack of knowledge on this particular topic, a specific task of the Italian Project Cometa-Autoctone Mediterranean crops and their valorization with advanced green chemistry technologies (funded by Ministry of Education, Universities and Research), was addressed to develop an effective handling and storage strategy for cardoon roots aimed to inulin production. In this framework the present study aimed to investigate the effect of chipping and drying approach on inulin content of cardoon roots biomass over 6 months of storage.

2. Materials and Methods

2.1. Location of the Field and Plant Material

Cardoon roots were taken in May 2020 from three year old plants cultivated in Terni (42.561335 N latitude, 12.62860 E longitude) (Umbria Region, Central Italy) on clay soil. Cardoon (cultivar Trinaseed) were grown from seeds sowed in November 2017 with a precision sowing machine at the rate of 3.0 kg/ha of seed and spacing distance of 0.75×0.17 m.

At the sowing, the seedbed was prepared between October and November with ploughing at 20 cm, followed by harrowing at 10–15 cm. During the first year, mineral fertilization with 46 kg of P2O5 and 64 kg/ha of N was added, and a.i. pendimethalin was used to control weeds. Starting from the second year, 46 kg/ha of P2O5 and 18 kg/ha of N were applied during the vegetative growth of the plants in autumn and in the early spring before the stem elongation phase. The fertilization rates were calculated on the basis of the soil fertility and crop nutrient uptake. Crop water requirements were satisfied by rain.

Approximately 500 plants were randomly uprooted using an excavator carefully avoiding damage to both canopy and root systems. The whole plants were put in sealed bags and carried to the laboratory of The Research Centre for Engineering and Agro-Food Processing of the Council for agricultural research and economics (CREA) in Monterotondo, Central Italy (42 10019″ N latitude 12 62066″ E longitude) for sampling. Firstly, the soil was removed from roots using a cold-water pressure washer. Afterwards, the plants were left to dry naturally for a few minutes. Then the roots were mechanically cut off from the plants.

The bulk density of roots was determined using a box with an internal volume of 0.0064 m^3, the value was reported as kg m^{-3}. The box was filled with roots and weighed with a KERN GmbH dynamometer (CH 50K50 model-range of measurements 50 kg and sensitivity 50 g). Three samples were taken for mean value. In 30 randomly chosen plants the fresh weight of canopy and roots were measured using a precision scale (Kern PCB 6000-0). The length of leaves and roots of this plants sample was measured with a ruler. Roots were further investigated by determining the moisture content according to [37].

2.2. Chipping, Drying and Storage

After cleaning, roots were divided into two groups (treatments) to monitor the inulin content in dried chipped roots (CRt) and dried whole roots (WRt) of cardoon in 6-months storage. CRt was obtained by selecting 15 kg of randomly chosen roots that were fresh chipped using an electric 2.0 kW bio-shredder (Zanon, mod. BIO 3). The particle size distribution (PSD) of the chipped material produced was analyzed according to [38].

Chips were collected and put into the oven for drying at 60 °C until they reached constant weight. Simultaneously, a further 15 kg of randomly chosen whole roots were collected and put in the oven for drying. At constant weight, the whole roots were removed to obtain WRt. The apparent bulk density was measured three times in CRt and WRt, respectively, according to [39] for mean value estimation.

Storage of CRt and WRt was performed outside the building, under a farm shed. Specifically, chips from CRt were collected in jute bags while the whole roots from WRt treatment were piled up as shown in Figure 1 after being labeled and weighed individually.

Figure 1. Storage under a farm shed of dried chipped roots (CRt) on the left side and dried whole roots (WRt) on the right side.

Monthly, approximately 200 g of chips from CRt and five roots from WRt were sampled and sent to ENEA laboratory for inulin determination. WRt roots were bioshredded before the shipment. A representative subsample from each treatment was kept for dry matter assessment.

2.3. Weather Data Monitoring

During the entire storage period of 6 months, the main weather-climatic parameters such as temperature, precipitation and air humidity were recorded with a weather station "DAVIS VANTAGE PRO 2" (Davis Instruments, 3465 Diablo Avenue, Hayward, CA 94545-2778, USA) located in the proximity of the storage site and connected to wireless net. Data are shown in Figure 2.

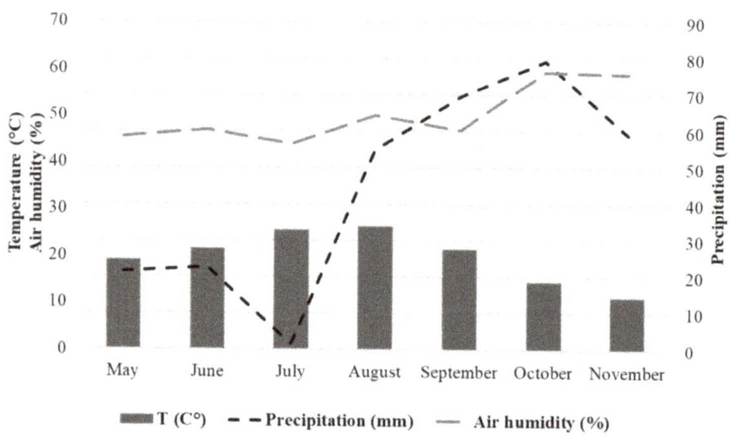

Figure 2. Trend of main climatic parameters: temperature (°C), precipitation (mm) and air humidity (%) recorded during the storage of cardoon roots from May to November 2020.

2.4. Inulin Content Determination

The collected samples from CRt and WRt were grinded to 0.5 mm in a ZM200 Retsch® ultracentrifugal mill (Retsch GmbH, Haan, Germany).

After elimination of residual humidity by drying at 50 °C for 4 h in a ventilated oven, the inulin content was determined by using a modified Raccuia method [19]. In particular, the quantitative extraction of the inulin from the roots was performed by suspending 1 g of powdered dry root in 20 mL of deionized water at 100 °C for 1 h in a Benchmark Scientific (South Plainfield, NJ, USA) Multi-Therm Heat-Shake kept under constant stirring at 500 rpm. Subsequently, the sample was centrifuged at 3500 rpm for 5 min and 4 mL of 0.75 M HCl were added to 2 mL of supernatant. The acid solution containing the inulin was then hydrolyzed in the heat-shake system for 15 min under the same conditions (100 °C, 500 rpm).

After centrifugation at 3500 rpm for 5 min, the supernatant was filtered through 0.45 μm PTFE filter (Whatman, USA) and carbohydrates were analyzed by using an HPIC DX 300 cromatographic system (Dionex, Sunnyvale, CA, USA) equipped with a Nucleogel® Ion 300 OA column (Macherey–Nagel, Düren, Germany) and sulphuric acid 10 mN as eluent. The detector was a Shodex RI101 refractive index (Showa Denko, Japan). All reagents and standards were purchased from Sigma-Aldrich (St. Louis, MO, USA). The extraction and hydrolysis processes were conducted in triplicate for each sample.

The inulin content was determined by the following Equation (1):

$$\text{Inulin }\% = \frac{\left(C_f + C_g\right) \times 0.9 \times 3}{C_R} \times 100 \quad (1)$$

where C_f and C_g are the concentration in gL^{-1} of fructose and glucose, respectively; 0.9 is the correction factor applied for the oligomer-to-monomer hydration; 3 is the dilution factor for the HCl hydrolysis; C_R is the concentration in gL^{-1} of the initial suspended roots.

2.5. Statistical Analysis

Statistical analysis was performed to assess significant differences among the mean values of dry matter and inulin content. Normality and homoscedasticity of the data were tested with Shapiro test and F test, respectively. *T*-test was performed to investigate significantly different means ($p \leq 0.05$) among treatments. Statistical analysis was performed by R 3.6.1 software to separate statistically different means [40].

3. Results and Discussion

3.1. Characterization of Cardoon Roots and Evaluation of Drying Times

The growth analysis of sample plants was performed in order to estimate the available aboveground and belowground biomass. The average fresh weight of canopy and roots was, respectively, 0.9 and 0.45 kg per plant (71 and 35 t f.w. ha^{-1}). The moisture content of roots was assessed as 70% *w/w* of fresh weight. Hence, the expected quantity of dry roots per hectare can be estimated in 10.6 t, similarly to 9.8 t DM ha^{-1} reported by [13]. Results of roots' characterization are given in Table 1.

Table 1. Characteristics of roots and canopy of three year old cardoon plants (mean ± sd) sampled in May 2020.

Taproot Length	Root Fresh Weight	Canopy Height	Canopy Fresh Weight
cm	kg	cm	kg
35.2 ± 14.9	0.45 ± 0.23	69.4 ± 6.6	0.9 ± 0.6

The bulk density of either fresh chips and dried chips was about 2.5 times higher than the bulk density of fresh whole roots and dried whole roots, respectively (Table 2).

Table 2. Bulk density of the whole roots and the chipped material, before and after drying

	Bulk Density (kg m^{-3})
Fresh whole roots	164.2 ± 15.9
Dried whole roots	61.79 ± 1.11
Fresh chips	418.8 ± 50.8
Dried chips	157.2 ± 31.6

Therefore, chipping represents an advisable option to reduce the volume needed for both transportation and storage. Consequently, the cost of the operations can be reduced as well. Although it was not investigated in the present study, sieving can follow the chipping phase to help removing unwanted debris from chips, which could be detrimental for further industrial processes. With this aim the chipping should be carried out with a forestry chipper able to produce a more homogeneous product.

On the contrary, in our case, as shown by PSD analysis (Figure 3), 90% of the chipped material was less than 8 mm length, making it not possible to separate by sieving unwanted debris from chips.

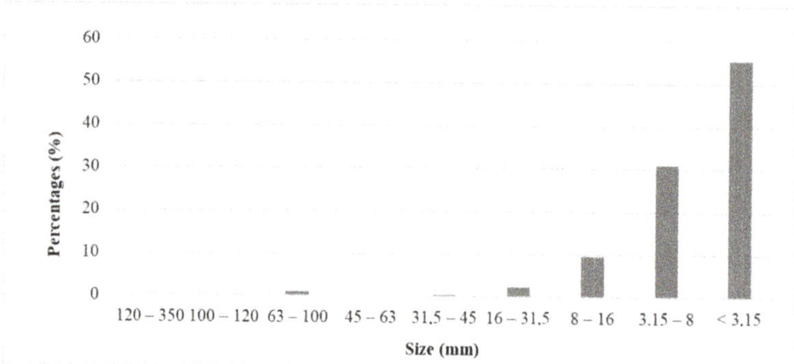

Figure 3. Particle size distribution of the chipped roots.

Moreover, the drying process can also benefit from chipping by reducing time and energy required [41]. According to our results, indeed, chipped roots could reach constant weight after 48 h, whilst whole roots needed 24 h more to dry completely (Figure 4).

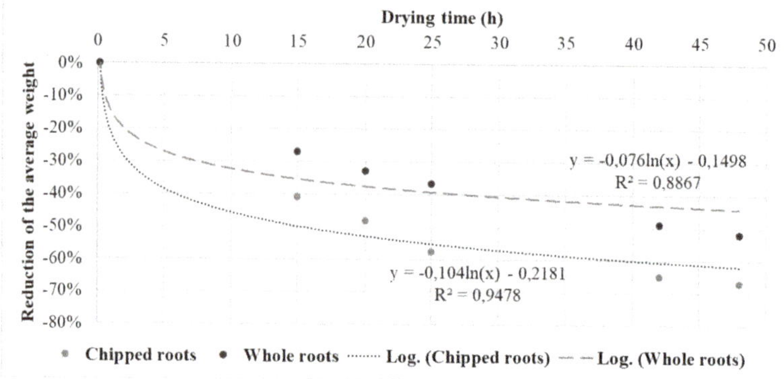

Figure 4. Reduction of the average weight of whole roots and chipped roots during drying in a thermo-ventilated oven at 60 °C.

3.2. Inulin Content

Inulin content at T0 (beginning of storage, immediately after drying) was 43.5 ± 0.65% and 47.1 ± 1.30% *w/w* in WRt and CRt, respectively (Figure 5). Drying time negatively affected the inulin content in WRt which resulted in 3.54% *w/w* lower than CRt. Conceivably, in WRt treatment, the inner tissues of the roots took longer to dry out than the outmost tissues of the same root. Hence, the metabolic activities naturally occurring in living cells stopped later and this partially explains the loss of inulin in WRt.

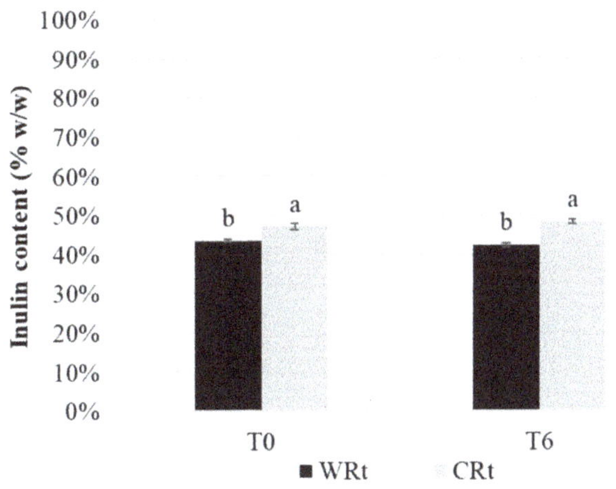

Figure 5. Inulin content in dry whole and chipped roots of cardoon at T0 and after 6 months of storage (T6). Common letters denote the absence of significant difference ($p < 0.05$).

Regardless of the difference in inulin content found among treatments, the values herein reported are consistent with the literature although some authors also highlight that possible changes in inulin content may be experienced according to the harvest season [14]. Higher values are usually found in spring, particularly between full blossom and fruit ripening. For this reason, root sampling was performed in May when the concentration is supposed to peak.

After 6 months storage (T6), inulin content did not change significantly in comparison with the initial content measured in the respective treatment, namely: 42.3 ± 0.82% *w/w* in WRt and 48.3 ± 1.12% *w/w* in CRt. Therefore, the drying process performed before the storage prevented the degradation of inulin, at least over the following 6 months of storage. This finding highlights the possibility for industries to exploit the drying process at the industrial scale to help storing the cardoon roots for longer time since inulin loss is prevented effectively. Artificial drying is certainly costly in terms of money and energy, but it is surely less costly than freezing. Additionally, the machinery required for drying is easier to run and cheaper to buy (e.g., a ventilated oven) with enormous advantage also for the transportation which does not longer require the ice-chain. Chipping also contributes to enhance the inulin supply chain as the higher bulk density of chipped material would require less space for drying, storage and transportation.

3.3. Moisture Content and Dry Matter Content

The moisture content at the first month of storage (T1) in WRt and CRt increased significantly up to 12.1% and 10.8%, respectively. This was probably due to the reabsorption of humidity from the external environment. In fact, as shown in Figure 6, the monthly moisture content measured in WRt and CRt traces the air moisture pattern recorded by weather station. The moisture increase in roots was more evident in WRt where 16.3% *w/w* of moisture was recorded at T6 (i.e., 3.4% higher than CRt). This was probably due to a greater exposure of the whole roots to air humidity with respect to the chipped material stored inside the jute bag.

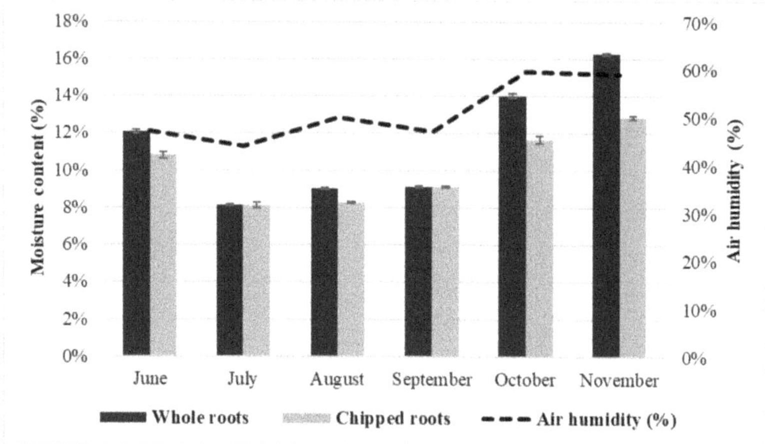

Figure 6. Moisture content (mean ± SD) in WRt and CRt during the 6 months storage in comparison with air moisture recorded by a weather station over the same period.

During the first month of storage, a significant reduction in weight in both treatments was recorded: 10.9% and 3.5% in WRt and CRt, respectively (Figure 7). Despite this, during the following months the roots' dry weight remained constant in both treatments. However, a significant difference of approximately 6% was recorded between the treatments throughout the trial. Both results were probably due to the higher water content recorded in the whole roots—not completely removed after drying (less drying efficiency) or greatly reabsorbed from the external environment (higher exposure to air humidity)—which promoted microbial activity.

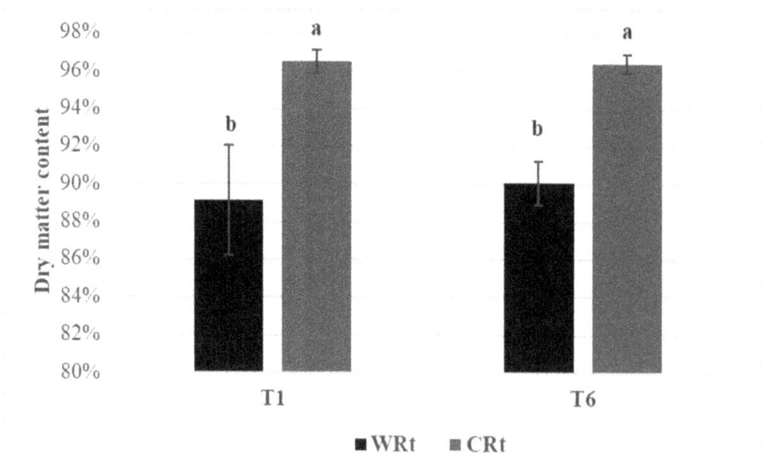

Figure 7. Dry matter content in dried whole roots (WRt) and dried chipped (CRt) roots at the first month of storage (June, T1) and after 6 months of storage (T6). Common letters denote the absence of significant difference ($p < 0.05$).

4. Conclusions

In the perspective of new generation biorefineries and the circular bioeconomy framework, the exploitation of cardoon also for inulin production is rather appealing, particularly if plants have been previously exploited for the production of further high added-value raw materials like seeds and stalks. Due the limited favorable period for harvesting the roots when inulin content is maximum, industries need to store enormous quantities of roots and process them gradually. Hence, storage plays a fundamental role in supply chain. Our findings suggest that during a 6-month storage inulin loss is negligible if roots are previously dried. Furthermore, chipping could also be a good practice since it is possible to reduce the volume required for storage (and also transportation) while it promotes a quicker drying; thus, less energy is required to dry out the roots.

In conclusion, our results highlight the possibility to chip cardoon roots meant for inulin extraction to ameliorate the supply chain of such a material. Although drying remains a costly strategy, chipping would help to reduce such cost by reducing the time required. However, further studies should provide clues to improve also the harvesting and cleaning process.

Author Contributions: Conceptualization, L.P. and V.A.; methodology, V.A., W.S. and F.L. (Francesco Latterini); validation, A.C.; formal analysis, V.A., W.S., F.L. (Francesco Latterini), F.L. (Federico Liuzzi) and V.V.; investigation, V.A., W.S., F.L. (Francesco Latterini), F.L. (Federico Liuzzi) and V.V.; resources, A.C.; writing—original draft preparation, V.A., F.L. (Federico Liuzzi), V.V.; writing—review and editing, V.A., W.S. and F.L. (Francesco Latterini).; supervision, L.P. and I.D.B.; funding acquisition, L.P. and I.D.B. All authors have read and agreed to the published version of the manuscript.

Funding: This research was performed within the Italian Project COMETA-Autoctone Mediterranean crops and their valorization with advanced green chemistry technologies. The project was funded by Ministry of Education, Universities and Research in the frame of PON "Ricerca e Innovazione" 2014–2020 and FSC/ARS01_00606 COMETA/CUP. B2G18000180004/Azione II.

Institutional Review Board Statement: Not applicable.

Informed Consent Statement: Informed consent was obtained from all subjects involved in the study.

Data Availability Statement: Data is not publicly available, though the data may be made available on request from the corresponding author.

Acknowledgments: The authors would like to thank Salvatore Antonino Raccuia for providing suggestions and sharing his knowledge on cardoon roots and storage methods.

Conflicts of Interest: The authors declare no conflict of interest.

References

1. Pappalardo, H.D.; Toscano, V.; Puglia, G.D.; Genovese, C.; Raccuia, S.A. *Cynara cardunculus* L. as a Multipurpose Crop for Plant Secondary Metabolites Production in Marginal Stressed Lands. *Front. Plant Sci.* **2020**, *11*, 240. [CrossRef] [PubMed]
2. Mauromicale, G.; Sortino, O.; Pesce, G.R.; Agnello, M.; Mauro, R.P. Suitability of cultivated and wild cardoon as a sustainable bioenergy crop for low input cultivation in low quality Mediterranean soils. *Ind. Crop. Prod.* **2014**, *57*, 82–89. [CrossRef]
3. Neri, U.; Pennelli, B.; Simonetti, G.; Francaviglia, R. Biomass partition and productive aptitude of wild and cultivated cardoon genotypes (*Cynara cardunculus* L.) in a marginal land of Central Italy. *Ind. Crop. Prod.* **2017**, *95*, 191–201. [CrossRef]
4. Ciancolini, A.; Alignan, M.; Pagnotta, M.A.; Vilarem, G.; Crinò, P. Selection of Italian cardoon genotypes as industrial crop for biomass and polyphenol production. *Ind. Crop. Prod.* **2013**, *51*, 145–151. [CrossRef]
5. Fernández, J.; Curt, M.D.; Aguado, P.L. Industrial applications of *Cynara cardunculus* L. for energy and other uses. *Ind. Crop. Prod.* **2006**, *24*, 222–229. [CrossRef]
6. Pari, L.; Alfano, V.; Mattei, P.; Santangelo, E. Pappi of cardoon (*Cynara cardunculus* L.): The use of wetting during the harvesting aimed at recovering for the biorefinery. *Ind. Crop. Prod.* **2017**, *108*, 722–728. [CrossRef]
7. Gominho, J.; Curt, M.D.; Lourenço, A.; Fernández, J.; Pereira, H. *Cynara cardunculus* L. as a biomass and multi-purpose crop: A review of 30 years of research. *Biomass Bioenergy* **2018**, *109*, 257–275. [CrossRef]
8. Caporusso, A.; De Bari, I.; Valerio, V.; Albergo, R.; Liuzzi, F. Conversion of cardoon crop residues into single cell oils by *Lipomyces tetrasporus* and *Cutaneotrichosporon curvatus*: Process optimizations to overcome the microbial inhibition of lignocellulosic hydrolysates. *Ind. Crop. Prod.* **2021**, *159*, 113030. [CrossRef]

9. Godard, A.; De Caro, P.; Thiebaud-Roux, S.; Vedrenne, E.; Mouloungui, Z. New Environmentally Friendly Oxidative Scission of Oleic Acid into Azelaic Acid and Pelargonic Acid. *J. Am. Oil Chem. Soc.* **2013**, *90*, 133–140. [CrossRef]
10. Petrone, M.T.; Stoppiello, G.; De Bari, I. Environmental impact of second-generation sugars production from cardoon residues. *Environ. Eng. Manag. J.* **2017**, *16*, 1769–1774. [CrossRef]
11. Cavalaglio, G.; Cotana, F.; Nicolini, A.; Coccia, V.; Petrozzi, A.; Formica, A.; Bertini, A. Characterization of Various Biomass Feedstock Suitable for Small-Scale Energy Plants as Preliminary Activity of Biocheaper Project. *Sustainability* **2020**, *12*, 6678. [CrossRef]
12. Ciria, C.; Sanz, M.; Carrasco, J.; Ciria, P. Identification of Arable Marginal Lands under Rainfed Conditions for Bioenergy Purposes in Spain. *Sustainability* **2019**, *11*, 1833. [CrossRef]
13. Raccuia, S.A.; Melilli, M.G. *Cynara cardunculus* L., a potential source of inulin in the Mediterranean environment: Screening of genetic variability. *Aust. J. Agric. Res.* **2004**, *55*, 693. [CrossRef]
14. Raccuia, S.A.; Melilli, M.G. Seasonal dynamics of biomass, inulin, and water-soluble sugars in roots of *Cynara cardunculus* L. *Field Crop. Res.* **2010**, *116*, 147–153. [CrossRef]
15. Christaki, E.; Bonos, E.; Florou-Paneri, P. Nutritional and Functional Properties of *Cynara* Crops (Globe Artichoke and Cardoon) and Their Potential Applications: A Review. *Int. J. Appl. Sci. Technol.* **2012**, *2*, 64–70.
16. Niness, K.R. Inulin and oligofructose: What are they? *J. Nutr.* **1999**, *129*, 1402–1406. [CrossRef]
17. López-Molina, D.; Navarro-Martínez, M.D.; Rojas-Melgarejo, F.; Hiner, A.N.P.; Chazarra, S.; Rodríguez-López, J.N. Molecular properties and prebiotic effect of inulin obtained from artichoke (*Cynara scolymus* L.). *Phytochemistry* **2005**, *66*, 1476–1484. [CrossRef]
18. Mensink, M.A.; Frijlink, H.W.; Van Der Voort Maarschalk, K.; Hinrichs, W.L.J. Inulin, a flexible oligosaccharide I: Review of its physicochemical characteristics. *Carbohydr. Polym.* **2015**, *130*, 405–419. [CrossRef]
19. Raccuia, S.A.; Genovese, C.; Leonardi, C.; Bognanni, R.; Platania, C.; Calderaro, P.; Melilli, M.G. Fructose production by *Cynara cardunculus* inulin hydrolysis. *Proc. Acta Hortic. Int. Soc. Hortic. Sci.* **2016**, *1147*, 309–314. [CrossRef]
20. Martínez, C.M.; Adamovic, T.; Cantero, D.A.; Cocero, M.J. Ultrafast hydrolysis of inulin in supercritical water: Fructooligosaccharides reaction pathway and Jerusalem artichoke valorization. *Ind. Crop. Prod.* **2019**, *133*, 72–78. [CrossRef]
21. Lv, S.; Wang, R.; Xiao, Y.; Li, F.; Mu, Y.; Lu, Y.; Gao, W.; Yang, B.; Kou, Y.; Zeng, J.; et al. Growth, yield formation, and inulin performance of a non-food energy crop, Jerusalem artichoke (*Helianthus tuberosus* L.), in a semi-arid area of China. *Ind. Crop. Prod.* **2019**, *134*, 71–79. [CrossRef]
22. Clauser, N.M.; González, G.; Mendieta, C.M.; Kruyeniski, J.; Area, M.C.; Vallejos, M.E. Biomass Waste as Sustainable Raw Material for Energy and Fuels. *Sustainability* **2021**, *13*, 794. [CrossRef]
23. Gholami, H.; Raouf Fard, F.; Saharkhiz, M.J.; Ghani, A. Yield and physicochemical properties of inulin obtained from Iranian chicory roots under vermicompost and humic acid treatments. *Ind. Crop. Prod.* **2018**, *123*, 610–616. [CrossRef]
24. Melilli, M.G.; Branca, F.; Sillitti, C.; Scandurra, S.; Calderaro, P.; Di Stefano, V. Germplasm evaluation to obtain inulin with high degree of polymerization in Mediterranean environment. *Nat. Prod. Res.* **2020**, *34*, 187–191. [CrossRef]
25. Pari, L.; Scarfone, A.; Santangelo, E.; Figorilli, S.; Crognale, S.; Petruccioli, M.; Suardi, A.; Gallucci, F.; Barontini, M. Alternative storage systems of *Arundo donax* L. and characterization of the stored biomass. *Ind. Crop. Prod.* **2015**, *75*, 59–65. [CrossRef]
26. Saengthongpinit, W.; Sajjaanantakul, T. Influence of harvest time and storage temperature on characteristics of inulin from Jerusalem artichoke (*Helianthus tuberosus* L.) tubers. *Postharvest Biol. Technol.* **2005**, *37*, 93–100. [CrossRef]
27. Maicaurkaew, S.; Jogloy, S.; Hamaker, B.R.; Ningsanond, S. Fructan:fructan 1-fructosyltransferase and inulin hydrolase activities relating to inulin and soluble sugars in Jerusalem artichoke (*Helianthus tuberosus* Linn.) tubers during storage. *J. Food Sci. Technol.* **2017**, *54*, 698–706. [CrossRef] [PubMed]
28. Pandino, G.; Barbagallo, R.N.; Lombardo, S.; Restuccia, C.; Muratore, G.; Licciardello, F.; Mazzaglia, A.; Ricceri, J.; Pesce, G.R.; Mauromicale, G. Quality traits of ready-to-use globe artichoke slices as affected by genotype, harvest time and storage time. Part I: Biochemical and physical aspects. *LWT Food Sci. Technol.* **2017**, *76*, 181–189. [CrossRef]
29. Löhmar, K.; Theurillat, V. Chicory beverages. In *Encyclopedia of Food Sciences and Nutrition*; Elsevier: Amsterdam, The Netherlands, 2003; pp. 1144–1149.
30. Towey, R.; Webster, K.; Darr, M. Influence of Storage Moisture and Temperature on Lignocellulosic Degradation. *AgriEngineering* **2019**, *1*, 332–342. [CrossRef]
31. Mujumdar, A.S.; Law, C.L. Drying Technology: Trends and Applications in Postharvest Processing. *Food Bioprocess Technol.* **2010**, *3*, 843–852. [CrossRef]
32. Chung, L.L.; Waje, S.S.; Thorat, B.N.; Mujumdar, A.S. Advances and recent developments in thermal drying for bio-origin and agricultural products. *Stewart Postharvest Rev.* **2008**, *4*, 1–23.
33. Suardi, A.; Latterini, F.; Alfano, V.; Palmieri, N.; Bergonzoli, S.; Pari, L. Analysis of the Work Productivity and Costs of a Stationary Chipper Applied to the Harvesting of Olive Tree Pruning for Bio-Energy Production. *Energies* **2020**, *13*, 1359. [CrossRef]
34. Singh, R.; Brar, G.S. Optimized biomass collection and transportation to biomass-based power plants. *J. Adv. Res. Dyn. Control Syst.* **2020**, *12*, 2322–2327.
35. Spinelli, R.; Marchi, E. Trends and Perspectives in the Design of Mobile Wood Chippers. *Croat. J. For. Eng.* **2021**, *42*, 25–38. [CrossRef]

36. Barontini, M.; Scarfone, A.; Spinelli, R.; Gallucci, F.; Santangelo, E.; Acampora, A.; Jirjis, R.; Civitarese, V.; Pari, L. Storage dynamics and fuel quality of poplar chips. *Biomass Bioenergy* **2014**, *62*, 17–25. [CrossRef]
37. ISO 18134-2:2017. *Solid Biofuels—Determination of Moisture Content—Oven Dry Method—Part 2: Total Moisture—Simplified Method*; ISO: Geneva, Switzerland, 2017; Available online: https://www.iso.org/standard/71536.html (accessed on 31 March 2021).
38. ISO 17827-1:2016. *Solid Biofuels—Determination of Particle Size Distribution for Uncompressed Fuels—Part 1: Oscillating Screen Method Using Sieves with Apertures of 3.15 mm and above*; ISO: Geneva, Switzerland, 2016; Available online: https://www.iso.org/obp/ui/#iso:std:iso:17827:-1:ed-1:v1:en (accessed on 31 March 2021).
39. ISO 17828:2015. *Solid Biofuels—Determination of Bulk Density*; ISO: Geneva, Switzerland, 2015; Available online: https://www.iso.org/obp/ui/#iso:std:iso:17828:ed-1:v1:en (accessed on 31 March 2021).
40. R Development Core Team. R: A Language and Environment for Statistical Computing. Available online: http://www.r-project.org (accessed on 6 March 2020).
41. Del Giudice, A.; Acampora, A.; Santangelo, E.; Pari, L.; Bergonzoli, S.; Guerriero, E.; Petracchini, F.; Torre, M.; Paolini, V.; Gallucci, F. Wood Chip Drying through the Using of a Mobile Rotary Dryer. *Energies* **2019**, *12*, 1590. [CrossRef]

sustainability

MDPI

Article

Environmental and Economic Assessment of Castor Oil Supply Chain: A Case Study

Luigi Pari, Alessandro Suardi *, Walter Stefanoni, Francesco Latterini and Nadia Palmieri

Consiglio per la Ricerca in Agricoltura e l'analisi dell'Economia Agraria (CREA)—Centro di Ricerca Ingegneria e Trasformazioni Agroalimentari, Via della Pascolare 16, Monterotondo, 00015 Rome, Italy; luigi.pari@crea.gov.it (L.P.); walter.stefanoni@crea.gov.it (W.S.); francesco.latterini@crea.gov.it (F.L.); nadia.palmieri@crea.gov.it (N.P.)

* Correspondence: alessandro.suardi@crea.gov.it; Tel.: +39-06-9067-5248

Received: 30 June 2020; Accepted: 5 August 2020; Published: 6 August 2020

Abstract: Among the species currently cultivated for industrial vegetable oil production, castor could be a good candidate for future investments due to the good resistance to pests, tolerance to drought, and suitability for marginal lands cultivation. In addition, the production of castor oil from Ricinus generates a large quantity of press cake, husks, and crop residues that, in a framework of bioeconomy, could be used as by-products for different purposes. Using a case study approach, the work presents results of the environmental impact assessment and economic feasibility of the production of castor oil from two different castor hybrids comparing four by-products management scenarios and two harvesting systems (manual vs. mechanical). Castor hybrid C-856 harvested manually and that involved only the soil incorporation of press cake obtained by the oil extraction resulted as the most sustainable. The hybrid C-1030 resulted as more profitable than C-856 when harvested with the combine harvester. The ratio between gross margin and GWP emissions was applied to calculate the economic performance (gross margin) per unit of environmental burden. Findings showed that Sc1B scenario in case of C-856 cultivar hybrid had a better ratio between economic performance and greenhouse gas (GHG) emitted into the atmosphere (€3.75 per kg CO_2eq).

Keywords: bioeconomy; life cycle assessment; life cycle costing; *Ricinus communis*, L.; castor oil; harvesting; by-product; residue management

1. Introduction

The world's population is estimated to exceed 9 billion people by 2050 according to FAO (2009). Thus, increasing in food and energy demand worldwide cannot be avoided: projections show that the overall food production is expected to rise by 70% [1] while the global demand for energy will increase by more than a quarter to 2040 [2]. Bioenergy production primarily aims at the greenhouse gas (GHG) reduction and achieving such a goal may lead to indirect land use change. Competition for land use among food and non-food crops is a serious issue that European Commission has been addressing for decades, and more stringent policy measures regarding sustainable production of food and energy are on the Agenda. On 1 January 2021 the proposed new directive RED II will enter into force, setting the new thresholds for minimum renewable energy share [3]. Investments on biofuel production from non-food feedstock are largely promoted by UE. New policy measures aim to achieve a 27% renewable energy share consumed by the electricity, heating and cooling, and transportation sectors by 2030 [3]. The adoption of energy crops could generate benefits from the reduction of fossil energy dependence, improvement of rural economies, and the achievement of environmental goals [4]. Biodiesel production from vegetable oils is feasible and widely accepted as an alternative strategy to meet these goals: It has similar properties to oil-derived diesel and, furthermore, it produces lower sulfur emission. Among the species currently cultivated for industrial vegetable oil production, castor could be a good candidate for future investments due to the good resistance to pests, tolerance to drought, and suitability for marginal lands cultivation [5]. According to FAO, in 2017, almost 1.8 million tons of castor seed had been produced worldwide, and Europe is the main user [6]. Furthermore, according to industry executives, the worldwide castor oil market is growing: The global castor oil market was $1180 million in 2018 and is expected to reach $1470 million by the end of 2025, growing at a compound annual growth rate (CAGR) of 2.8 percent between 2019 and 2025, according to international reports [7]. The price of castor oil in the beginning of 2019 in the international market reached 1600 dollars per ton compared with 1300 dollars per ton of 2018 [8].

In addition, the production of castor oil from Ricinus generates a large quantity of press cakes, husks, and crop residues [9] that, in a framework of bioeconomy, could be used as by-products for different purposes. In this framework, the European Project MAGIC (Marginal lands for Growing Industrial Crops—Grant Agreement number: 727698-MAGIC-H2020-RUR-2016-2017/H2020-RUR-2016-2) aims towards the development of resource-efficient and economically profitable industrial crops to be grown on marginal land. Among industrial crops considered in the Project, there is *Ricinus communis*, L. (castor) that is cultivated for its seed oil, which is employed extensively in medicine, pharmaceuticals, and biorefineries [10]. Castor is a vigorous fast-growing herbaceous plant native to tropical Africa [11,12] which is tolerant to salinity and drought stresses, with additional benefits of providing a multi-purpose oilseed production [13]. In the world, the most productive country is India (more than 80% of the worldwide production) along with Mozambique, China, Brazil, Myanmar, Ethiopia, Paraguay, and Vietnam. These are all developing countries that benefit from low labor costs, and the economic impact of the harvesting phase is thus sustainable. The lack of possibility to harvest the seeds mechanically is dictated by a high amount of aboveground biomass produced by wild cultivars that cannot be processed by common combine harvesters. In fact, clogging may occur in the case of a high quantity of aerial biomass production, and high seed losses. In order to solve this problem, breeders around the world are struggling to produce hybrids of castor exhibiting high productivity but shorter in height and with homogeneous ripening of the capsules.

To our knowledge, limited studies have been dedicated to the environmental and economic sustainability of castor [14,15]. Some studies are focused on the sustainability assessment of the residue biomass utilization [16] or biodiesel production [17] without investigating in detail the impact of the various castor agricultural stages and different residue management. In particular, a comparison the of the environmental sustainability between different castor hybrids, harvesting methods, and by-products management have not been presented in the literature. Using a case study approach [18,19], the work aimed to present results concerning the estimation of the environmental impacts caused by two different castor hybrids harvested both manually and mechanically (manual vs. mechanical harvestings). Both hybrids had similar seed yields, even though hybrid C-856 is shorter than hybrid C-1030. The latter reported a higher amount of epigeal biomass. Various scenarios of on-farm by-products managements were analyzed. Starting by the same approach, the study carried out an economic assessment to identify the most advantageous scenario for each castor variety and residue management.

2. Materials and Methods

2.1. Study Sites

The study area is located in Geaca Municipality, Cluj District (Romania). Cluj District lies in the northwestern half of the country, between parallels 47°28′ in North and 46°24′ in South, meridians 23°39′ in west and 24°13′ in east, respectively. It is located in the contact zone of three representative natural units: Apuseni Mountains, Someș Plateau, and Transylvanian Plain. Cluj District is the 12th largest in the country and accounts for almost 3% of Romania's area. It is bordered to the northeast by Maramureș and Bistrița-Năsăud counties, to the east by Mureș District, to the south by Alba District, and to the west by Bihor and Sălaj counties.

The trials were carried out on the 8th and 9th of October 2019 in two different experimental fields where castor was harvested (Figure 1).

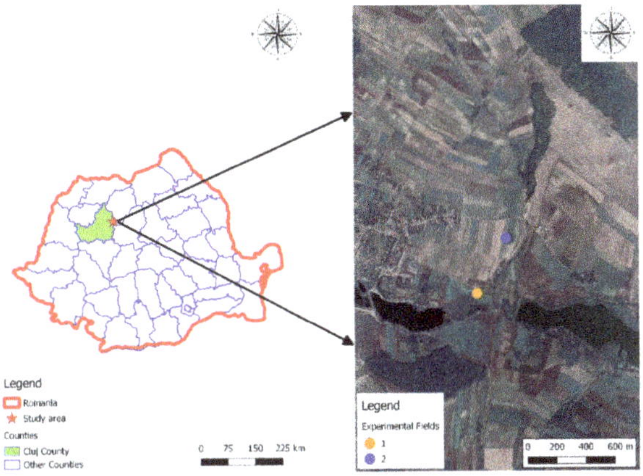

Figure 1. Study area and experimental field location (Geaca Municipality, Cluj District, Romania).

Main features of the experimental fields are given in Table 1. Data were taken both from GIS analysis and from field relieves with clinometer.

As highlighted in Table 1, all fields have southern exposition and the prevalent altitude of 313 m a.s.l. was recorded in Field 2 while the maximum slope was recorded in Field 1. However, both fields can be considered flat terrains. The surface of the field 2 was 0.27 ha higher than Field 1. A view of experimental fields positioning on Sentinel-2 image dated 3 September 2019 is given in Figure 2.

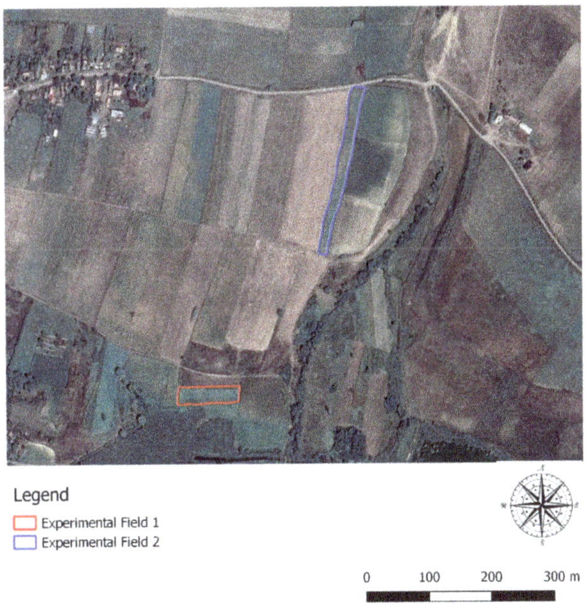

Figure 2. Experimental fields positioning. Base map Google Satellite Images dated 3 September 2019.

Table 1. Main features of the experimental fields for castor harvesting.

Experimental Field	Prevalent Slope [%]	Minimum Slope [%]	Maximum Slope [%]	Prevalent Exposition	Prevalent Altitude [m a.s.l.]	Surface [ha]
1	8.5	7.9	8.7	South	294	0.25
2	5.7	4.1	7.2	South	313	0.47

2.2. *Crop Characteristics and Management of the by-Products*

The main data of two dwarf hybrids of *Ricinus communis* (C-856 and C-1030) collected during the trials are reported in Table 2. Plants were cultivated in Romania, and seeds were provided to local farmer (Ecoricinus—National association of Ricinus growers) by the Israelian company KAIIMA.

Table 2. Primary data: Two castor hybrids. Pre-harvest data collection: Height of the plants, aerial biomass produced, and Harvest Index.

Hybrid Cultivar	Height of Plants [cm]	Husks [Mg ha^{-1}]	Seed [Mg ha^{-1}]	Straw Fresh Weight [Mg ha^{-1}]		Harvest Index [%]
		d.w.	ssf.w.	f.w.	d.w.	
C-856	74.4 c	1.40	2.80 a	4.13 b	0.87 b	52.5 a
C-1030	112.8 a	1.60	2.90 a	8.35 a	1.61 a	43.4 b

Note: Common letters within columns denote the absence of significant difference ($p < 0.05$).

The dwarf hybrids tested were two of the various chosen by the association Ecoricinus to evaluate their behavior and productivity in Romania. Although hybrid C-856 has already been analyzed in productive and morphological terms by Alexopoulou et al. [6] in Greece and Italy, hybrid C-1030 has never been described in the literature and in the present study, it has been analyzed only to assess its productivity and the amount of epigeal biomass available for the LCA study.

Despite the significant difference found in height, straw production, and the harvest index (HI) between the two hybrids, in both cases, the aboveground biomass produced was lower than the quantity produced by wild varieties commonly cultivated in Romania (data not shown). Therefore, more suitable for mechanical harvesting. The farmer reported that fertilization and plowing took place in 2018 between week 47 and 48, while harrowing and sowing occurred in week 23 and 24 2019 at the depth of 8–10 cm with the sowing density of 3.6 seeds m^{-2}. Mature cow manure was applied at the quantity of 6 Mg ha^{-1} and no irrigation was provided. No chemicals were used for both weed control and desiccation of leaves.

On the basis of farmer's survey, two scenarios for each castor variety with a different mix of by-products management were considered (Table 3).

Table 3. Scenarios analyzed in the study for each castor variety (Sc1 = scenario 1; Sc2 = scenario 2).

Harvesting Systems		Manual Harvesting		Mechanical Harvesting	
Scenarios		Sc1A	Sc1B	Sc2A	Sc2B
Products and co-products	Straw	Soil incorporation	Sale	Soil incorporation	Sale
	Husk	Sale	Sale	Soil incorporation	Soil incorporation
	Press cake	Soil incorporation	Soil incorporation	Soil incorporation	Soil incorporation
	Castor oil	Sale	Sale	Sale	Sale

In the case of manual harvesting (Sc1A and Sc1B), the castor fruit is harvested as whole, and the separation of the spiny capsules from the seeds takes place on the farm. According to Parascanu (2017), castor husk might be deemed as the best candidate for the combustion process due to its high heat release [20]. Therefore, in the Sc1A and Sc1B scenarios, the sale of spiny capsules has been assumed at a market price of crushed olive stones due to similar lower heating value (LHV) that results in 16.48 [20] and16.50 MJ/kg [21], respectively.

In the scenarios Sc2A and Sc2B involving castor mechanical harvesting, castor husk has always been considered incorporated into the soil because it was discharged on the ground by the combine harvester as residue and not collected. Castor straw is a residue with an LHV of 17.68 MJ/kg and an ash content of 1.70 wt% [20]. For this reason, both manual and mechanical harvesting scenarios have been considered, both sold as solid biofuel (Sc1B and Sc2B) and incorporated into the soil (Sc1A and Sc2A). According to the farmer, press cake that resulted during the oil extraction phase is used as fertilizer and for this reason was always considered incorporated into the soil. Castor oil is the main product in the supply chain, and it has always been considered as sold.

2.3. Data Sampling and Measurements

Pre-harvest tests were conducted directly in the field. Four plots of 1.5m x 2m each were randomly selected within the two experimental fields in order to measure the growth of the plants and estimate the aboveground biomass produced. In each plot, plants were counted and cut at the collect level, then brought outside the field for height measurements as well as straw and capsules fresh weight determination. The height of the plant was taken by measuring the distance between the collect and the tip of the longest raceme. Samples of straw and the total capsules collected in each plot were put in sealed bags and brought to the laboratory for dry weight determination. In the laboratory, capsules were separated manually from the seeds. Thus, seeds were weighed for seed yield estimation. Simultaneously, samples of straw and empty capsules (husks) were dried at constant temperature of 105 °C in a ventilated oven until constant weight was reached (EN ISO 18134-2:2015). Then, the dry matter and humidity content were calculated. All data were subjected to the analysis of variance (ANOVA) to separate statistically different means ($P < 0.05$).

2.4. Life Cycle Assessment of Castor Oil Supply Chain

An environmental impact analysis of castor oil production was carried out using the life cycle assessment methodology (LCA) according to UNI EN ISO 14040: 2006 [22] and UNI EN ISO 14044: 2006 [23], by means an attributional approach [24–27], including the following statements: (a) Goal definition and scoping: Defining the goals of the study, the functional units, the boundaries of the system, and the required data; (b) life cycle inventory: data collection; (c) life cycle impact assessment: Estimation of the potential environmental impacts; (d) life cycle interpretation and improvement: Final step where the risks are evaluated and checked to draw conclusions.

2.4.1. Goal Definition and Scoping

The considered system is defined by all the agricultural processes that occurred during the *Ricinus communis* growing phase and subsequent oil extraction phase carried out at farm level.

The boundary of the system (Figure 3) is given by the life cycle stages of castor to be included in the LCA. Cultivation phases and extraction phase of oil were studied from cradle-to-farm gate.

The functional unit represents the reference unit used to quantify all inputs and outputs from the boundaries of the system. It is defined as 1 Mg of castor oil produced by the farm.

Firstly, the environmental impact of each single hybrid cultivar was separately analyzed for each scenario; then, each scenario was assessed to identify the best hybrid cultivar.

Allocation describes how environmental impacts are shared between the main product and co-products along the supply chain [28]. Castor oil is the main product, while crop residues (castor husks and straw) and press cakes are considered co-products [9]. In an LCA study, the co-product handling is a crucial issue because it could impact on the final results [29]. Agricultural products are particularly sensitive to allocation methods because of the different share that their co-products can have. In our case, an economic allocation method that takes into account market prices and mass of product and by-products per each scenario was used [30] (Table 4). Castor market prices are not easy to be find, especially for castor co-products that do not have a market. For this reason, the selling price of castor seed was considered to be 600 euros per Mg, while the price of castor oil for cosmetic purposes was 30 euros per liter according to informal local market. As described above, in the absence of a market, husk and straw prices for energy purposes have been assimilated to solid biomass with similar characteristics (olive stones and wheat straw) used for energy purposes and with known market prices. In fact, following information from the informal local market as happen in other studies [31], the price of husks for energy purposes was 180 euros per Mg, while the price of straw was considered 55 euros per Mg. The prices used are those indicated by the Ricinum producers National Association of Romania (Ecoricinus Productie Comert Srl, Cluj-Napoca, 10, Fanatelor st. jud, Cluj, Romania). Even if the press cake corresponds to an important amount of biomass, due to its returns to the soil as fertilizer internally at the farm, according to the economic allocation type used and due to an absence of a market and a market price, the impact generated by the press cake was assumed to be very low (0.07%) with a minimum price of 0.1 € per Mg of by-product.

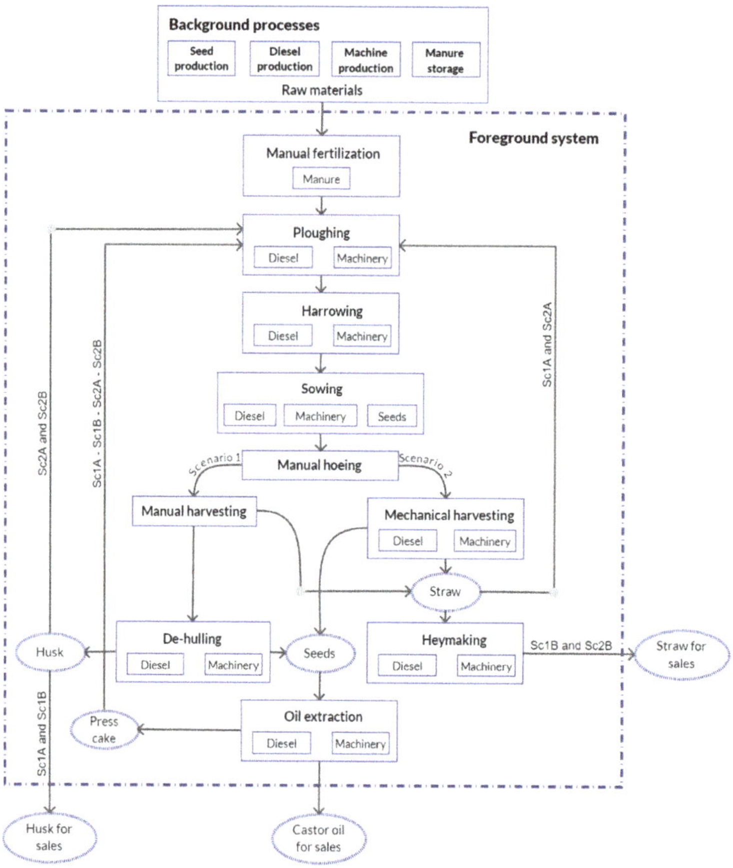

Figure 3. System boundary.

Table 4. Economic allocation factors for each castor variety.

Phases	Product and by-Products	Cultivar Hybrids	
		C-856	C-1030
Agricultural phases	Husks with seed	97.50%	95.84%
	Straw	2.50%	4.16%
Total		100.00%	100.00%
Dehulling	Castor seed	86.96%	85.80%
	Husks	13.04%	14.20%
Total		100.00%	100.00%
Oil extraction phase	Castor oil	99.93%	99.93%
	Press cake	0.07%	0.07%
Total		100.00%	100.00%

Source: CREA elaboration.

2.4.2. Life Cycle Inventory Analysis

Data resulting from a survey carried out by field technicians were utilized for the life cycle inventory analysis. The Simapro code database 8.0.2 (Prè Consultants, Amersfoort, The Netherlands) was used for data not identifiable by survey.

The primary data were relative to the technical characteristics of the tractors and agricultural equipment utilized and diesel consumption (Table 5). Regarding the hypothesis of mechanical harvesting, all data for the costs, performance, and specifications of a conventional combine harvester were derived from personal communication and literature. Moreover, the primary data were relative to different castor varieties.

Table 5. Technical characteristics of the machineries, diesel consumption, and agricultural phases.

Agricultural Operation	Manual Fert.	Plough.	Harr.	Sow.	Manual Hoeing	Manual Harv.*	Mech Harv.**	Dehull.	Oil Extrac.
Machinery									
Machinery power (kW)	-	78	78	78	-	-	146	7.7	3
Machinery weight (kg)	-	3750	3750	3750	-	-	10700	250	1900
Fuel consumption ($l\ ha^{-1}$)	-	45	15	5	-	-	25	5.1	-
Lubrificant consumption ($l\ ha^{-1}$)		0.10	0.07	0.03	-	-	0.05	0.18	-
Lifetime ($h\ ha^{-1}$)	-	12,000	12,000	12,000	-	-	3000	2000	-
Instrument used (type)	Shovel	Moldboard plow	Rolmako	Row planter	-	-	-	-	-
Instrument power (kW)		66	63	44	-	-	-	-	-
Weight instruments (kg)	1.5	795	2860	830	-	-	-	-	-
Lifetime (h)	4000	2000	2000	1500	-	-	-	-	-
Product utilized (type)	Manure	-	-	Seeds	-	-	-	-	-
Quantity ($kg\ ha^{-1}$)	6000	-	-	15	-	-	-	-	-

Source: CREA elaboration on survey data. *Scenario 1. ** Scenario 2.

The secondary data referred to the emission generated by the machines during different agricultural phases and from fertilizers.

Emissions in air, soil, and underground water (leaching) due to manure storage, as well as by-products and manure incorporation into the soil per each scenario, were calculated using the model proposed by [32] and values of the references reported in Table 6.

Table 6. Secondary data: Source of the emissions considered in the study for storage and soil incorporation of the manure, and by-products (press cake, straw, and husk according to the scenario—Table 2).

Emissions	Source
Manure storage emissions	
Emissions of methane (CH_4) and nitrous oxide (N_2O)	[33–35]
Ammonia (NH_3) emissions due to manure storage	[35,36]
Nitrogen oxides (NOx) emissions	[37], using the factor by [34]
Emissions related to soil incorporation different combinations of by-products	
CO_2 emissions	[6,38]
N_2O emissions	[32]
Emissions due to soil incorporation of manure	
N_2O, NH_3, NOx and nitrate leaching	[32]
Emission factor of Potassium, Copper and Zinc	[36]

The exhaust gases emissions from agricultural tractors and combine harvester were calculated using the standard emission factors for diesel engines reported by Directive 2004/26/EC for carbon-nitrogen oxides (g NOx ha^{-1}), hydrocarbons (g HC ha^{-1}), monoxide (g CO ha^{-1}), and particulate matter (g PM ha^{-1}), according to the method reported by [39]. The amount of released carbon dioxide (kg CO_2 ha^{-1}) was calculated by multiplying the fuel consumption (kg ha^{-1}) by an air emission factor of 2.6 (kg CO_2 emitted per kg of diesel fuel consumed), according to [40,41].

2.4.3. Land Use Change (LUC)

The direct and indirect land use change (LUC) associated with crop production can produce changes in the carbon from soil and vegetation [42]. Castor oil can be in the form of herbaceous or arborescent plant, annual or perennial, depending on the climatic conditions of the region. In the present study, castor oil is cultivated as annual oil crop in cropland that had not undergone any land-use conversion for a period of more than 20 years [34]. Following the indications of the Intergovernmental Panel on Climate [34] there is no net accumulation of biomass carbon stocks for annual crops. On the other hand, emission from soil carbon mineralization per each scenario has been taken into consideration because there are changes in the management activities on croplands, and in particular, in the amount of biomass that has been considered incorporated into the soil according to the different scenario considered (Table 2). Even if the soil's organic carbon was considered in the steady state, and the farm analyzed employed crop rotations, different crop residue management considered in the study and the amount of GHG emitted during the different scenario were calculated according to the following formula:

$$GHG_{res} = \sum_{i=1}^{3} \left(Res_i \times C_{res_i} \times C_{min_i} \times aw_{CO_2} \right) \tag{1}$$

where

$GHGres$ = Greenhouse gases emissions from soil incorporation of residue "i" per scenario (Mg CO_2 ha^{-1})
Res_i = Amount of residue "i" incorporated into the soil (Mg ha^{-1})
$Cres_i$ = Organic carbon content in the residues "i" (%) [6]
$Cmin_i$ = Organic carsbon in the residues "i" mineralized in soil (%) [38]
$awCO_2$ = atomic weight of carbon dioxide equal to 44/12

2.4.4. Life Cycle Impact Assessment

The environmental impact of 1 Mg of castor oil was based on GHG emissions. The carbon footprint was defined as the sum of all GHGs emitted within the system boundary and expressed in CO_2 equivalent applying the IPCC 2007 method (100-year life span).

A parallel economic assessment is integrated with LCA also using a life cycle perspective that covers all activities in the supply chain up to the farm gate. The economic sustainability is critical because when it comes to assessing the different products and by-products management, the attention of farmers does not fall solely on environmental impacts, but also (and mainly) on economic aspects. For this reason, an economic assessment was carried out.

2.5. Economic Assessment

The study followed the steps in LCA identified in the relevant international standard [22,23] with the corresponding steps in life cycle costing (LCC) introduced in parallel. Life cycle costing (LCC) is a methodology that. aimed to assess the costs across the entire life cycle of a product, process, or service [43] concentrating on the economic cost at each stage [44]. A conventional cradle-to-gate LCC was applied here and includes the assessment of all costs associated with the life cycle of the castor-oil cultivation specific to each scenario. In particular, the LCC assessment is focused on internal costs (value of goods and services consumed, including raw materials, services, other operating expenses, and labor costs). It is important to underline that the contractors provide all phases of the preparation of the field up to sowing (bottom fertilization, ploughing, harrowing, and sowing). Everything afterwards (weed control and harvesting) is performed by the owners of the field for Sc1A and Sc1B. In Sc2A and Sc2B, all agricultural phases are in subcontractor account. Later, to evaluate the gross margin of farm, the revenues for each product (multiplying between prices and quantity of products) are calculated. Gross margin refers to the difference between revenue from crop sales and the variable costs related to the agricultural activities [44] and it is a profitability indicator of a farm. All data (Table 7) come from the budget (year 2018) of the farm studied.

Table 7. Economic data expressed in €/ha per year.

	Cultivar Hybrids	
Costs (€/Year)	**C-856**	**C-1030**
Manual fertilization	200.00	200.00
Ploughing	120.00	120.00
Harrowing	60.00	60.00
Sowing	60.00	60.00
Manual hoeing	375.00	375.00
Manual harvesting	625.00	625.00
Mechanical harvesting	180.00	180.00
Dehulling	150.00	150.00
Oil extraction	390.00	390.00
Revenues (€/year)		
Straw for sales	49.5	88.00
Husks for sales	255.00	288.00
Castor oil for sales	26,206.32	27,142.26

Source: CREA elaboration.

3. Results and Discussions

According to the literature, castor yield can change appreciably with genotype [45]. Arnaud (1990) observed a seed yield from 2000 to 2620 kg ha^{-1} in France [46], while Anastasi (2015) reported a yield between 1790 to 4750 kg ha^{-1} in Italy [47]. In the present research, the genotypes of castor grown showed similar productions of 2800 and 2900 kg ha^{-1} for C-856 and C-1030, respectively.

However, the C-1030 hybrid, which is higher than C-856 and has a significantly higher HI (Table 2), produced 85% more straw than C-856, with the same inputs used.

Alexopoulou et al. [6], from the comparison of various castor hybrids planted in Greece and Italy, found an average amount of stems and leaves of 1.08 Mg_{dm} ha^{-1}, and the hybrid C-856, that resulted as 133 cm tall (79% taller than in our study) in Greece (Aliartos area, Greece in 2014), allowed for obtaining 1.13 Mg_{dm} ha^{-1} of stems and leaves against 0.87 Mg ha^{-1} obtained in the present study. In the same study, the C-856 hybrid produced a straw quantity of 0.585 Mg ha^{-1}, much more similar to that obtained in this study in 2012 in Greece (Aliartos area, Greece in 2014) [6]. In general, Alexopoulou et al. [6] highlighted that C-856 resulted as the best-performing hybrid in Italy while in Greece, its yields were quite low, probably related to the high percentage of immature racemes (60%) at harvest. This suggests the influence of the climate and crop management on the phenotype expression of this hybrid. To the best of our knowledge, there is no information in the literature about the C-1030 hybrid.

The type of harvesting represents a critical phase that can also have a significant influence on the amount of product that can be collected per unit area. Mechanized harvesting allows for collecting about 3 t/h of castor oil seeds (considering a harvesting rate between 0.75 and 1.5 hectares per hour) ready to be pressed. On the other hand, according to farmers, manual harvesting shows extremely low losses <5%. On the contrary, castor mechanized harvesting needs to be improved due to the major losses, which can be up to 50% as evidenced by [48]. So far, only one machine manufacturer has started the first harvesting tests using a specific castor header, which would be able to reduce losses to 5% [48], and Zhao et al. [49] reported the possibility to harvest the capsules using a vibrating system instead of a cutting bar [49]. In the present study, losses were not considered given the uncertainty of the data to be scientifically verified in specific tests.

3.1. LCA

The impact analysis allowed for identifying the processes that had higher impacts on the environment.

What emerged from the analysis was that fertilization was the agricultural phase with the most impact. This result is common to various studies [50–56]. In the present study, for all cultivar hybrids and scenarios, the environmental impacts of fertilization phase were due to emissions of methane (CH_4), dinitrogen monoxide (N_2O), and carbon dioxide (CO_2) from manure management and its incorporation into the soil. In fact, fertilization emitted 74 to 89% of the GHG of the castor oil production. The LCA study of biodiesel production from rapeseed published by Malça et al. [57] reported that the cultivation stage impacted 66 to 79% and fertilization was the main cause of GHG emissions [57]. According to our results, the higher GHG emissions were mainly due to the characteristics, and the direct and indirect emissions were generated by manure itself. It should be highlighted that, as suggested by Aguilera et al. [58], organic fertilizers applied at similar N rates to synthetic fertilizers generally make smaller contributions to the leached NO_3^- pool, and can mitigate N_2O emissions [58]. The different by-product management also influenced the indirect emissions of GHG due to their degradations during soil incorporation.

In the case of castor oil produced by both C-856 and C-1030 cultivar hybrids, as expected, the manual harvesting resulted as more sustainable (Sc1A and Sc1B), and Sc1B scenario was always the least impactful, followed by scenarios 1A and 2B (Figures 4 and 5).

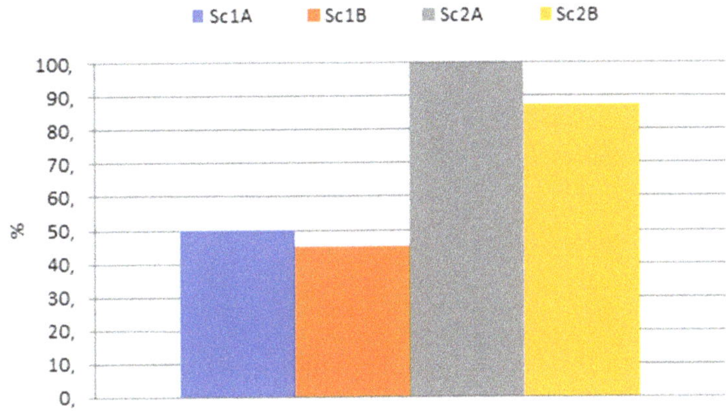

Figure 4. Carbon footprint of 1 Mg of castor oil hybrid C-856 for each scenario ((Sc1 = scenario 1; Sc2 = scenario 2).

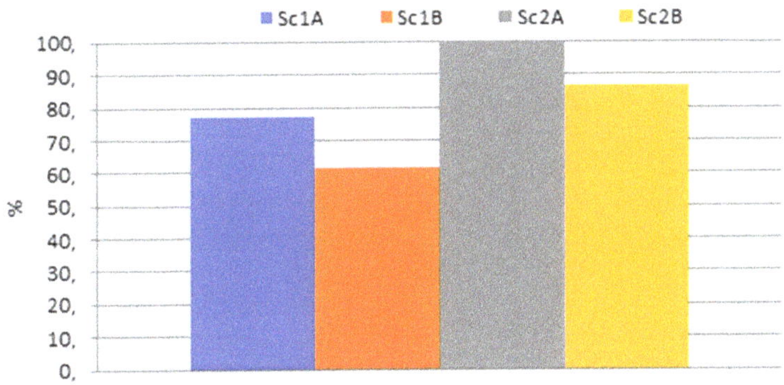

Figure 5. Carbon footprint of 1 Mg of castor oil hybrid C-1030 for each scenario (Sc1 = scenario 1; Sc2 = scenario 2).

Moreover, among cultivar hybrids and all scenarios, Sc2A_C-1030 is more impactful than the other treatments analyzed, while the Sc1B_C-856 is less burdensome than others. These results were due to both different combinations of on-farm by-products (castor press cake incorporation into the soil in case of 1B_C-856, and castor press cake, straw and husks incorporation into the soil in case of 2A_C-1030) and yields (2.8 Mg per ha in case of C-856 vs. 2.9 Mg per ha in case of C-1030). In general, the incorporation of by-products in the soil at farm level has resulted in higher GHG emissions than their sale. For this reason, the highest impact observed in the mechanized harvesting treatments (Sc2A and Sc2B) is largely due to the non-collection of husks that are left in the field by the combine and then buried (unlike manual harvesting where husks are separated from the seeds on the farm and then sold as solid fuel). Obviously, the study focused on the impacts related to the production of castor oil on the farm, not considering the whole process downstream of the supply chain and the related impacts that could completely reverse the results obtained.

The life cycle of scenario 1B, in which manual harvesting was assumed (less burdensome in the case of hybrid C-856, slightly less productive, and with less press cake), and with the incorporation of the pressed cake alone and the sale of the other by-products, resulted in the emission of 8.14 Mg CO_2eq per Mg of castor oil (8.14 kg CO_2eq per kg of castor oil extracted). On the other hand, the life cycle of scenario 2A_C-1030, in which mechanized harvesting with combine harvesters and the incorporation of straw, husks, and press cakes was assumed, resulted in the emission of 18.9 Mg CO_2eq per Mg of castor oil produced (18.9 kg CO_2eq per kg of castor oil extracted).

Although, in Sc1A and Sc1B scenarios, there is the de-hulling phase that there is not in Sc2A and Sc2B, this has a very small impact always <8% (on average 0.698 Mg CO_2eq Mg^{-1} of castor oil produced) of the total CO_2 emissions. The oil extraction impacted less than 5% of the total CO_2 emitted (on average 0.412 Mg CO_2eq Mg^{-1} of castor oil). Sanz Requena (2010) reported that for each ton of crude sunflower, rapeseed, and soybean oil extracted, an average of 2.2 Mg of CO_2 was emitted, but it should be highlighted that after the mechanical extraction, a treatment with a solvent (hexane) was included [54].

Spinelli (2012) reported a total emission of 13.7 Mg CO_2eq Mg^{-1} of sunflower oil produced [59]. However, according to the study and the allocation used, the emissions became 4.52 Mg CO_2eq Mg^{-1} of sunflower oil and 9.18 Mg CO_2eq Mg^{-1} of sunflower cake produced. The lack of allocation of a higher share of emissions from the press cake makes castor oil production inevitably more impactful in GHG emitted than other vegetable oils, although the variety C-856 with manual harvesting have relatively low and promising overall emissions.

3.2. Economic Assessment

The economic gross margin is related mainly to the yield level (product and by-products) and to the cost of inputs for each scenario and cultivar hybrid. As far as the yield is concerned, the values for each farm and crop have been previously discussed and the data are reported in Table 8, showing higher yields per ha for C-1030 than for C-856 crops. For these reasons, the C-1030 cultivar shows lower total costs per Mg cultivated than C-856 ones (Table 8). Moreover, for both cultivar hybrids, the total costs of manual harvesting scenario are higher than mechanical harvesting scenario ones. This finding was due to labor costs in harvesting phase. In fact, in case of the manual harvesting scenario, five workers are required to harvest castor seed, contributing 32% to the total costs; while in case of mechanical scenario one worker (with machinery) is required contributing just 13% to the total costs. The impact that manual harvesting has on costs can be equated to that reported by Silalertruksa (2012) in Thailand, where manual harvesting accounts for 22% of total costs in the palm oil sector [60].

Table 8. Economic gross margin for each scenario expressed in €/FU (1 Mg of castor oil)—(Sc1 = scenario 1; Sc2 = scenario 2).

	Cultivar Hybrid: C-856 Scenarios				**Cultivar Hybrid: C-1030 Scenarios**			
	Manual		**Mechanical**		**Manual**		**Mechanical**	
	Sc1A	*Sc1B*	*Sc2A*	*Sc2B*	*Sc1A*	*Sc1B*	*Sc2A*	*Sc2B*
Costs (€/Mg) *								
Manual fertilization	72.42	71.42	71.42	71.42	68.96	68.96	68.96	68.96
Ploughing	42.85	42.85	42.85	42.85	41.38	41.38	41.38	41.38
Harrowing	21.43	21.43	21.43	21.43	20.69	20.69	20.69	20.69
Sowing	21.43	21.43	21.43	21.43	20.69	20.69	20.69	20.69
Manual hoeing	133.93	133.93	133.93	133.93	129.31	129.31	129.31	129.31
Harvesting	223.21	223,21	64.28	64.28	215.52	215.52	62.07	62.07
Dehulling	53.57	53.57	-	-	51.72	51.72	-	-
Oil extraction	139.28	139.28	139.28	139.28	134.48	134.48	134.48	134.48
Total Costs (€/Mg)	708.12	708.12	494.62	494.62	682.75	682.75	477.58	477.58
Revenues (€/Mg)								
Straw for sales	-	4,58	-	4,58	-	7.61	-	7.61
Husks for sales	76.24	76.24	-	-	81.57	81.57	-	-
Castor oil for sales	31,166	31,166	31,166	31,166	31,166	31,166	31,166	31,166
Total Revenues (€/Mg)	31,242	31,246	31,166	31,170	31,247	31,255	31,166	31,173
Gross Margin (€/Mg)	30,533	30,537	30,671	30,675	30,564	30,572	30,688	30,695

Source: CREA elaboration on budget data (year 2018). * For each agricultural phase are included the internal costs (i.e., value of goods and services consumed, including raw materials, services, other operating expenses and labor costs).

Table 9 shows that the 2B_C-1030 scenario had higher gross margin than other scenarios; while the 1A_C-856 scenario had the lowest gross margin.

Table 9. Gross margin and carbon footprint for each scenario, expressed in €/FU (1 Mg of castor oil)—(Sc1 = scenario 1; Sc2 = scenario 2).

		Cultivar Hybrid: C-856 Scenarios				Cultivar Hybrid: C-1030 Scenarios			
	Unit	Manual		Mechanical		Manual		Mechanical	
		Sc1A	Sc1B	Sc2A	Sc2B	Sc1A	Sc1B	Sc2A	Sc2B
Gross Margin	**(€/FU)**	30,533	30,537	30,671	30,675	30,564	30,572	30,688	30,695
GWP	**(kg CO_2 eq/FU)**	9070	8140	18,100	15,800	14,600	11,600	18,900	16,300
Gross Margin/GWP ratio	**(%)**	3.37	3.75	1.69	1.94	2.09	2.63	1.62	1.88

Source: CREA elaboration on both budget data (year 2018) and environmental findings.

In addition, the ratio between gross margin and GWP emissions was applied to calculate the economic performance (gross margin) per unit of environmental burden (Table 9). The ratio is based on data from both environmental and economic accounting systems. The higher the ratio value, the higher the economic performance per unit of GWP emitted.

Findings showed that scenario 1B in the case of C-856 cultivar hybrid had a better ratio between economic performance and GHG emitted into the atmosphere (€3.75 per kg CO_2eq); while the 2A_C-1030 scenario showed the worst ratio between economic and environmental performances (€1.62 per kg CO_2-eq) confirming the environmental results. These results were due to different combinations of on-farm by-products (see Table 3), different revenues (see Table 8), and yields (see Table 2).

4. Conclusions

There has been a critical increment in interest for sustainable and biodegradable items so as to diminish reliance on petrochemicals. This is one of the essential elements which is driving the growth of the worldwide castor oil market. The research focused on the evaluation of the environmental and economic sustainability of two different castor hybrids (C-856 and C-1030) comparing manual and mechanical harvesting methods, and by-product management.

Comparing all the proposed scenarios, the cultivation of the manually harvested castor hybrid C-856 and the by-product management that involved only the soil incorporation of press cake obtained by the oil extraction resulted as the most sustainable. On the other hand, the mechanized harvesting of hybrid C-1030, which involved the incorporation of all the by-products of the cultivation of castor and production of castor oil (husk, straw, and press cake) showed the highest CO_2 emissions per Mg of castor oil (+132%). It is therefore clear how, with the same inputs used, the castor-oil cultivation method affects the management of by-products and how, while residues are a source of organic matter for the soil, they cause greenhouse gas emissions during the degradation process in the soil.

From an economic point of view, a difference in Gross Margin (€/Mg) between the hybrids used was only evident when comparing the scenarios in which mechanized harvesting was used, i.e., C-856_Sc2A vs. C-1030_Sc2A and C-856_Sc2B vs. C-1030_Sc2B, resulting in an increase in Gross Margin of 6 and 7%, respectively, using the hybrid C-1030. The two hybrids when harvested manually did not show appreciable increases in Gross Margin (0.1%). In general, the scenario that produced most Gross Margin was the C-1030_Sc2B where mechanized harvesting of the plants, the incorporation of husk and press cake, and the sale of castor oil and straw were carried out.

In the end, to determine the most economically and environmentally convenient scenario, the ratio between gross margin and GWP emissions was applied to calculate the economic performance (gross margin) per unit of environmental burden. Findings showed that scenario Sc1B in the case of C-856 cultivar hybrid had a better ratio between economic performance and GHG emitted into the atmosphere

(€3.75 per kg CO_2eq); while the Sc2A_C-1030 scenario showed the worst ratio between economic and environmental performances (€1.62 per kg CO_2eq) confirming the environmental results.

Although Sc1B represents a good economic–environmental compromise, including manual harvesting, it clashes both with the need to innovate the castor production chain, and with the costs and availability of labor that may vary over time, affecting the sustainability of the chain, costs, and market prices.

Furthermore, an important aspect that was not considered in the study is the loss of product during harvesting. This is particularly relevant in the case of very high losses that are reflected in the impacts per unit of product. With the implementation of well-functioning mechanized castor harvesting systems, the resulting seed losses will also necessarily have to be considered in future studies.

Moreover, it is important to highlight that the study did not consider the whole process downstream of the castor oil extraction and the related impacts that could completely reverse the results obtained, which should be investigated in future researches.

Ultimately, the lack of official economic data on the market prices of products and by-products, and the difficulty of finding the costs resulting from the various cultivation practices, within the castor production chain, as old as it is, currently undergoing improvement and remodernization, represents a limit to obtaining exhaustive answers on its economic sustainability. For this reason, this research does not have the presumption to provide a definitive answer to the questions related to the environmental and economic sustainability of the castor-oil production chain, which will need further study and analysis as the production methods are refined.

Author Contributions: Conceptualization and methodology, A.S., N.P., L.P.; investigation and data curation N.P., A.S., W.S., F.L.; writing—original draft preparation N.P., W.S., F.L., A.S.; writing—review and editing, A.S.; supervision, L.P. and A.S.; funding acquisition, L.P. All authors have read and agreed to the published version of the manuscript.

Funding: This research was funded by European Union's Horizon 2020 Magic—Marginal lands for Growing Industrial Crops: Turning a burden into an opportunity project grant number 727698 (http://https://magic-h2020.eu). The funders had no role in the design of the study; in the collection, analyses, or interpretation of data; in the writing of the manuscript, or in the decision to publish the results.

Acknowledgments: Authors thank the ricinum producers National Association of Romania (Ecoricinus Productie Comert Srl, Cluj-Napoca, 10, Fanatelor st. jud, Cluj, Romania) and his team for their valid support and assistance provided during the activities, the field data and data for the economic analysis, as well as Sandu Lazar for his valuable contribution in the field activities.

Conflicts of Interest: The authors declare no conflict of interest.

References

1. FAO (Food and Agriculture Organization). Global agriculture towards 2050. In Proceedings of the High Level Expert Forum—How to Feed the World, Rome, Italy, 12–13 October 2009; Volume 2050, pp. 1–4.
2. IEA (International Energy Agency). *World Energy Outlook—Executive Summary*; IEA: Paris, France, 2018; Volume 11.
3. ICCT (Internationa Council on Clean Transportation). *The European Commission's Renewable Energy Proposal for 2030*; ICCT: Berlin, Germany, 2017; Volume 8.
4. Zafeiriou, E.; Karelakis, C. Income volatility of energy crops: The case of rapeseed. *J. Clean. Prod.* **2016**, *122*, 113–120. [CrossRef]
5. Makanju, A.; Bello, E.I. Production, characterization and evaluation of castor oil biodiesel as alternative fuel for diesel engines. *J. Emerg. Trends Eng. Appl. Sci.* **2011**, *2*, 525–530.
6. Alexopoulou, E.; Papatheohari, Y.; Zanetti, F.; Tsiotas, K.; Papamichael, I.; Christou, M.; Namatov, I.; Monti, A. Comparative studies on several castor (Ricinus communis L.) hybrids: Growth, yields, seed oil and biomass characterization. *Ind. Crops Prod.* **2015**, *75*, 8–13. [CrossRef]
7. Murphy, C.B. Compound Annual Growth Rate—CAGR. Available online: https://www.investopedia.com/terms/c/cagr.asp (accessed on 22 June 2020).

8. Ghosal, S. Castor Oil Prices Spike 23% in Global Market. Available online: https://economictimes.indiatimes.com/markets/commodities/news/castor-oil-prices-spike-23-in-global-market/articleshow/69089709.cms (accessed on 22 June 2020).
9. Bateni, H.; Karimi, K.; Zamani, A.; Benakashani, F. Castor plant for biodiesel, biogas, and ethanol production with a biorefinery processing perspective. *Appl. Energy* **2014**, *136*, 14–22. [CrossRef]
10. Mensah, M.B.; Awudza, J.A.M.; O'Brien, P. Castor oil: A suitable green source of capping agent for nanoparticle syntheses and facile surface functionalization. *R. Soc. Open Sci.* **2018**, *5*, 180824. [CrossRef]
11. Huang, H.; Yu, N.; Wang, L.; Gupta, D.K.; He, Z.; Wang, K.; Zhu, Z.; Yan, X.; Li, T.; Yang, X. The phytoremediation potential of bioenergy crop Ricinus communis for DDTs and cadmium co-contaminated soil. *Bioresour. Technol.* **2011**, *102*, 11034–11038. [CrossRef]
12. Rajkumar, M.; Freitas, H. Influence of metal resistant-plant growth-promoting bacteria on the growth of Ricinus communis in soil contaminated with heavy metals. *Chemosphere* **2008**, *71*, 834–842. [CrossRef]
13. Pandey, V.C. Suitability of Ricinus communis L. cultivation for phytoremediation of fly ash disposal sites. *Ecol. Eng.* **2013**, *57*, 336–341. [CrossRef]
14. Helling, R.K.; Russell, D.A. Use of life cycle assessment to characterize the environmental impacts of polyol production options. *Green Chem.* **2009**, *11*, 380–389. [CrossRef]
15. Amouri, M.; Mohellebi, F.; Zaïd, T.A.; Aziza, M. Sustainability assessment of Ricinus communis biodiesel using LCA approach. *Clean Technol. Environ. Policy* **2017**, *19*, 749–760. [CrossRef]
16. Parascanu, M.M.; Puig-Gamero, M.; Soreanu, G.; Valverde, J.L.; Sanchez-Silva, L. Comparison of three Mexican biomasses valorization through combustion and gasification: Environmental and economic analysis. *Energy* **2019**, *189*, 116095. [CrossRef]
17. Liang, S.; Xu, M.; Zhang, T. Life cycle assessment of biodiesel production in China. *Bioresour. Technol.* **2013**, *129*, 72–77. [CrossRef] [PubMed]
18. Adewale, C.; Reganold, J.P.; Higgins, S.; Evans, R.D.; Carpenter-Boggs, L. Agricultural carbon footprint is farm specific: Case study of two organic farms. *J. Clean. Prod.* **2019**, *229*, 795–805. [CrossRef]
19. Mittal, J.P.; Dhawan, K.C.; Thyagraj, C.R. Energy scenario of castor crop under dryland agriculture of Andhra Pradesh. *Energy Convers. Manag.* **1991**, *32*, 425–430. [CrossRef]
20. Parascanu, M.M.; Sandoval-Salas, F.; Soreanu, G.; Valverde, J.L.; Sanchez-Silva, L. Valorization of Mexican biomasses through pyrolysis, combustion and gasification processes. *Renew. Sustain. Energy Rev.* **2017**, *71*, 509–522. [CrossRef]
21. Perea-Moreno, M.-A.; Manzano-Agugliaro, F.; Hernandez-Escobedo, Q.; Perea-Moreno, A.-J. Peanut shell for energy: Properties and its potential to respect the environment. *Sustainability* **2018**, *10*, 3254. [CrossRef]
22. ISO (International Organization for Standardization). *Environmental Management—Life Cycle Assessment—Principles and Framework*; ISO 14040:2006; ISO: Geneva, Switzerland, 2006.
23. ISO (International Organization for Standardization). *Environmental Management—Life Cycle Assessment—Requirements and Guidelines*; ISO 14044:2006; ISO: Geneva, Switzerland, 2006.
24. Palmieri, N.; Forleo, M.B.; Giannoccaro, G.; Suardi, A.; Pari, L. Environmental and economic performance of cereal straw end-practices. In Proceedings of the 25th European Biomass Conference and Exhibition, Stockholm, Sweden, 12–15 June 2017; Volume 2017.
25. Ekvall, T.; Azapagic, A.; Finnveden, G.; Rydberg, T.; Weidema, B.P.; Zamagni, A. Attributional and consequential LCA in the ILCD handbook. *Int. J. Life Cycle Assess.* **2016**, *21*, 293–296. [CrossRef]
26. Finnveden, G.; Hauschild, M.Z.; Ekvall, T.; Guinée, J.; Heijungs, R.; Hellweg, S.; Koehler, A.; Pennington, D.; Suh, S. Recent developments in Life Cycle Assessment. *J. Environ. Manag.* **2009**, *91*, 1–21. [CrossRef]
27. McManus, M.C.; Taylor, C.M. The changing nature of life cycle assessment. *Biomass Bioenergy* **2015**, *82*, 13–26. [CrossRef]
28. De Boer, I.J.M. Environmental impact assessment of conventional and organic milk production. *Livest. Prod. Sci.* **2003**, *80*, 69–77. [CrossRef]
29. Notarnicola, B.; Salomone, R.; Petti, L.; Renzulli, P.A.; Roma, R.; Cerutti, A.K. *Life Cycle Assessment in the Agri-Food Sector: Case Studies, Methodological Issues and Best Practices*; Springer: Berlin/Heidelberg, Germany, 2015; ISBN 3319119400.
30. Chen, C.; Habert, G.; Bouzidi, Y.; Jullien, A.; Ventura, A. LCA allocation procedure used as an incitative method for waste recycling: An application to mineral additions in concrete. *Resour. Conserv. Recycl.* **2010**, *54*, 1231–1240. [CrossRef]

31. Vásquez Lavin, F.; Barrientos, M.; Castillo, Á.; Herrera, I.; Ponce Oliva, R.D. Firewood certification programs: Key attributes and policy implications. *Energy Policy* **2020**, *137*, 111160. [CrossRef]
32. Brentrup, F.; Kusters, J.; Lammel, J.; Kuhlmann, H. Methods to estimate on-field nitrogen emissions from crop production as an input to LCA studies in the agricultural sector. *Int. J. Life Cycle Assess.* **2000**, *5*, 349–357. [CrossRef]
33. Dalla Riva, A.; Kristensen, T.; De Marchi, M.; Kargo, M.; Jensen, J.; Cassandro, M. Carbon footprint from dairy farming system: Comparison between Holstein and Jersey cattle in Italian circumstances. *Acta Agrar. Kvár.* **2014**, *18*, 75–80.
34. Eggleston, H.S.; Buendia, L.; Miwa, K.; Ngara, T.; Tanabe, K. *IPCC Guidelines for National Greenhouse Gas Inventories*; Agriculture, Forestry and Other Land Use; IPCC: Hayama, Japan, 2006; Volume 4.
35. ISPRA (Istituto Superiore per la Protezione e la Ricerca Ambientale). *Agricoltura—Inventario Nazionale delle Emissioni e Disaggregazione Provinciale (Agriculture—National Emission Inventory and Provincial Breakdown)*; Report 85/2008; ISPRA: Rome, Italy, 2008; ISBN 9788844805012.
36. Falconi, F.; Neri, P.; Borsari, A.; Bombardieri, R.; Di Stefano, M.; Brambilla, C.; Querzola, F. Analisi Ambientale del Ciclo di Vita della Produzione di Latte da Allevamento Biologico e Confronto con la Convenzionale. In *ENEA e UTS Protezione e Sviluppo dell'Ambiente e del Territorio*; ENEA: Rome, Italy, 2005.
37. Battini, F.; Agostini, A.; Tabaglio, V.; Amaducci, S. Environmental impacts of different dairy farming systems in the Po Valley. *J. Clean. Prod.* **2016**, *112*, 91–102. [CrossRef]
38. Ruíz-Valdiviezo, V.M.; Mendoza-Urbina, L.D.; Luna-Guido, M.; Gutiérrez-Miceli, F.A.; Cárdenas-Aquino, M.R.; Montes-Molina, J.A.; Dendooven, L. Emission of CO_2, CH_4 and N_2O and dynamics of mineral N in soils amended with castor bean (Ricinus communis L.) and piñón (Jatropha curcas L.) seed cake. *Plant Soil Environ.* **2013**, *59*, 51–56.
39. Forleo, M.B.; Palmieri, N.; Suardi, A.; Coaloa, D.; Pari, L. The eco-efficiency of rapeseed and sunflower cultivation in Italy. Joining environmental and economic assessment. *J. Clean. Prod.* **2017**, *172*, 3138–3153. [CrossRef]
40. Lal, R.; Stewart, B.A. *Soil-Specific Farming: Precision Agriculture*; Advances in Soil Science; CRC Press: Boca Raton, FL, USA, 2015; ISBN 9781482245349.
41. Grace, P.R.; Harrington, L.; Jain, M.; Robertson, G.P. Long-Term sustainability of the tropical and subtropical rice-wheat system: An environmental perspective. *Improv. Product. Sustain. Rice Wheat Syst. Issues Impacts* **2003**, *65*, 27–43.
42. Iriarte, A.; Rieradevall, J.; Gabarrell, X. Life cycle assessment of sunflower and rapeseed as energy crops under Chilean conditions. *J. Clean. Prod.* **2010**, *18*, 336–345. [CrossRef]
43. Rigamonti, L.; Borghi, G.; Martignon, G.; Grosso, M. Life cycle costing of energy recovery from solid recovered fuel produced in MBT plants in Italy. *Waste Manag.* **2019**, *99*, 154–162. [CrossRef]
44. Brandão, M.; Clift, R.; Milà, L.C.; Basson, L. A life-cycle approach to characterising environmental and economic impacts of multifunctional land-use systems: An integrated assessment in the UK. *Sustainability* **2010**, *2*, 3747–3776. [CrossRef]
45. Koutroubas, S.D.; Papakosta, D.K.; Doitsinis, A. Adaptation and yielding ability of castor plant (Ricinus communis L.) genotypes in a Mediterranean climate. *Eur. J. Agron.* **1999**, *11*, 227–237. [CrossRef]
46. Arnaud, F. The development of castor-oil crops in France. In *Il Ricino: Obiettivi, Strategie e Ricerca. Agricoltura Ricerca*; Ministero Agricoltura e Foreste Roma: Rome, Italy, 1990.
47. Anastasi, U.; Sortino, O.; Cosentino, S.L.; Patanè, C. Seed yield and oil quality of perennial castor bean in a Mediterranean environment. *Int. J. Plant Prod.* **2015**, *9*, 99–116.
48. Pari, L.; Latterini, F.; Stefanoni, W. Herbaceous Oil crops, a review on mechanical harvesting state of the art. *Agriculture* **2020**, *10*, 309. [CrossRef]
49. Zhao, H.; Zhang, C. Analysis on the research status and structure characteristics of castor harvester. In Proceedings of the 2019 IEEE International Conference on Mechatronics and Automation (ICMA), Tianjin, China, 4–7 August 2019; pp. 415–420.
50. Palmieri, N.; Forleo, M.B.; Giannoccaro, G.; Suardi, A. Environmental impact of cereal straw management: An on-farm assessment. *J. Clean. Prod.* **2017**, *142*, 2950–2964. [CrossRef]
51. Pari, L.; Palmieri, N.; Forleo, M.B.; Suardi, A.; Coaloa, D.; Friuli, V.G. Lca of oilseed rape production for energy purposes: Sensitivity analysis. In Proceedings of the 20th European Biomass Conference and Exhibition, Milan, Italy, 18–22 June 2012.

52. Li, S.; Huang, B.; Zhao, F.; Lu, Z.; Wang, X.; Chen, X.; Chen, Y. Environmental impact assessment of agricultural production in Chongming ecological island. *Int. J. Life Cycle Assess.* **2019**, *24*, 1937–1947. [CrossRef]
53. Chatzisymeon, E.; Foteinis, S.; Borthwick, A.G.L. Life cycle assessment of the environmental performance of conventional and organic methods of open field pepper cultivation system. *Int. J. Life Cycle Assess.* **2017**, *22*, 896–908. [CrossRef]
54. Sanz Requena, J.F.; Guimaraes, A.C.; Quirós Alpera, S.; Relea Gangas, E.; Hernandez-Navarro, S.; Navas Gracia, L.M.; Martin-Gil, J.; Fresneda Cuesta, H. Life Cycle Assessment (LCA) of the biofuel production process from sunflower oil, rapeseed oil and soybean oil. *Fuel Process. Technol.* **2011**, *92*, 190–199. [CrossRef]
55. Cherubini, F.; Bird, N.D.; Cowie, A.; Jungmeier, G.; Schlamadinger, B.; Woess-Gallasch, S. Energy- and greenhouse gas-based LCA of biofuel and bioenergy systems: Key issues, ranges and recommendations. *Resour. Conserv. Recycl.* **2009**, *53*, 434–447. [CrossRef]
56. Adler, P.R.; Del Grosso, S.J.; Parton, W.J. Life cycle assessment of net greenhouse gas flux for bioenergy cropping systems. *Ecol. Appl.* **2007**, *17*, 675–691. [CrossRef]
57. Malça, J.; Coelho, A.; Freire, F. Environmental life-cycle assessment of rapeseed-based biodiesel: Alternative cultivation systems and locations. *Appl. Energy* **2014**, *114*, 837–844. [CrossRef]
58. Aguilera, E.; Lassaletta, L.; Sanz-Cobena, A.; Garnier, J.; Vallejo, A. The potential of organic fertilizers and water management to reduce N_2O emissions in Mediterranean climate cropping systems. A review. *Agric. Ecosyst. Environ.* **2013**, *164*, 32–52. [CrossRef]
59. Spinelli, D.; Jez, S.; Basosi, R. Integrated Environmental Assessment of sunflower oil production. *Process Biochem.* **2012**, *47*, 1595–1602. [CrossRef]
60. Silalertruksa, T.; Bonnet, S.; Gheewala, S.H. Life cycle costing and externalities of palm oil biodiesel in Thailand. *J. Clean. Prod.* **2012**, *28*, 225–232. [CrossRef]

sustainability

Article

Economic and Environmental Assessment of Two Different Rain Water Harvesting Systems for Agriculture

Luigi Pari, Alessandro Suardi, Walter Stefanoni, Francesco Latterini and Nadia Palmieri *

Consiglio per la Ricerca in Agricoltura e l'analisi dell'Economia Agraria (CREA)—Centro di Ricerca Ingegneria e Trasformazioni Agroalimentari, Via della Pascolare 16, Monterotondo, 00015 Rome, Italy; luigi.pari@crea.gov.it (L.P.); alessandro.suardi@crea.gov.it (A.S.); walter.stefanoni@crea.gov.it (W.S.); francesco.latterini@crea.gov.it (F.L.)

* Correspondence: nadia.palmieri@crea.gov.it; Tel.: +39-069-067-5219

Abstract: Increasing aridity and subsequent water scarcity are currently among the major problems of agriculture. Rainwater harvesting could represent a way to tackle this issue, and, as a consequence, scientific research has been more and more focused on such topic. On the other hand, few scientific studies related to economic and environmental assessment of rainwater harvesting systems in agriculture are available. The present study carried out an economic and environmental analysis of two different systems for rainwater harvesting: a typical pond and an innovative flexible water storage system (FWSS). The environmental and economic performance of the systems was compared using the Life Cycle Assessment (LCA) and Life Cycle Costing (LCC) methodologies, referring to a functional unit (FU) of 1 m^3 of storable water. The FWSS showed better environmental end economic performance than the pond system, resulting with both lower environmental impacts (17.04 g per m^3 CO_2 *vs* 28.2 g per m^3 CO_2) and lower costs (16.94 € per m^3 *vs* 20.41 € per m^3). Moreover, the pond system was more impactful than the FWSS for all the 17 categories investigated. Therefore, the FWSS can be a suitable solution for water harvesting in agriculture sector, showing interesting features for farmers.

Keywords: ecoefficiency; life cycle assessment (LCA); life cycle costing (LCC); run-off; pond; flexible water storage system

Citation: Pari, L.; Suardi, A.; Stefanoni, W.; Latterini, F.; Palmieri, N. Economic and Environmental Assessment of Two Different Rain Water Harvesting Systems for Agriculture. *Sustainability* **2021**, *13*, 3871. https://doi.org/10.3390/su13073871

Academic Editor: Hossein Azadi

Received: 10 March 2021
Accepted: 26 March 2021
Published: 31 March 2021

1. Introduction

Water scarcity and water supply are among the major concerns that countries worldwide have been struggling to address during the last decades. Usually, European countries are not arid, but some, like Cyprus, Bulgaria, Belgium, Spain, Malta and Italy, are currently exploiting 20% or more of their long-term water supplies every year. Agriculture is among the main responsible sector for freshwater consumption accounting for the 24% of the abstracted water that can go up to 80% in southern regions [1]. The need to rely on natural fresh water basins or on underground water is further fostered by the effect of climate change on the rainfall pattern in the Mediterranean region, where heavy rainfall events are occurring more frequently and only in a limited period of the year [2]. Farmers struggle to plan field activities, and plants need to be irrigated artificially more often than before. They mainly rely on underground water, but the overexploitation of such resource has detrimental effects on the environment [3]. Public awareness of agriculture impact on the environment is driving the change from conventional farming to organic farming, the latter of which seeks to burden the environment with as little water depletion as possible [4]. Although organic farming also contributes to GHGs reduction, it is not the resolutive strategy to cope with this problem. Interest is growing in arid and semi-arid regions of the planet concerning the possibility of collecting and storing rainwater for urban and agriculture purposes [5].

The present water scarcity leads to a new paradigm in water resource management, and the application of sustainable water supply solutions is essential [6,7]. Some studies [8,9] have catalyzed interest in alternative approaches to ensure water security by applying, for example, rainwater harvesting systems.

Rain water harvesting (RWH) is the process of accumulating incident raindrops on ground surfaces and roofs, with the help of cisterns, tanks, and underground check dams [10]. RWH has been applied for centuries by humankind to meet water supply needs and nowadays can still represent an important practice to improve the efficiency of the use of water in the urban future [5] by reducing household expenditures on water consumption [11] and offering many opportunities for agriculture.

In fact, irrigation of rainfed crops through RWH could represent a good option to increase crop yields due to an improvement in the water productivity [5].

Jiang et al. (2013) highlighted that the rainwater supplementary irrigation could increase crop yield by more than 30% [12]. Furthermore, even if rainfed agriculture in arid and semi-arid areas represents to up to 90% of the total production of cereal of these regions, in many countries, productivity remains low due to the sub-optimal rainfall characteristics, disadvantageous land conditions, and a deficiency in good management of these resources. On the other hand, increasing productivity of rainfed areas could lead to an increase in food security, reduction in irrigation frequency, and improvement in livelihoods and rural conditions. Furthermore, as reported by Ghimire et al. (2017) [13], RWH could reduce "impacts on the environment and human health, stormwater runoff and combined sewer overflows, and economic viability". As observed by various authors, RWH represents a valid system to reduce stormwater runoff, improving water management in an affordable manner [14–16]. Surface runoff is a phenomenon that, at the farm level, is triggered by 10 to 25% of rainwater falling in arid and semi-arid areas, and it can have negative effects on soil erosion and the accumulation of nutrients, chemicals, and sediments into rivers and streams [5]. Storage can be achieved with various types of storage systems that can also differ remarkably in terms of costs and environmental impacts [5].

Nevertheless, the environmental sustainability of different strategies to store rainwater has been seldomly investigated at the farm level; particularly, it has not been taken into account in the decision-making phase which should include this aspect along with the economical and feasibility aspects. Only few studies dealt with the environmental aspect of fresh water storage system, and, if so, they mainly focused on drinkable water [17,18]. Because agriculture is a highly demanding activity in fresh water (more than a simple beverage), much attention is ought to be paid to such aspect. Life Cycle Analysis (LCA) is widely recognized as a standardized method [19,20] that is used to evaluate the potential environmental impacts of products, processes, or services during the entire life cycle. Similar to LCA, Life Cycle Costing (LCC) methodology [21,22] is one of the main tools used to embed economic factors into the assessment of sustainability.

In general, as noted by several authors, irrigation in agriculture also leads to increased environmental impacts [23–26]. Most of the works in the literature focus on the emissions generated by the irrigation phase, focusing mainly on the amount of water resources used or on the energy related to the irrigation phase [23,24], and often without specifically analyzing the infrastructure (irrigation system) used [25,26].

On the other hand, some studies assessed the sustainability of rain water supply systems. Yan et al. (2018) [9] compared the environmental impacts of decentralized and centralized potable water supply, and they found a water-saving efficiency laying between 0.6 and 100%, depending on rainfall. Their results suggested that potable water produced from this decentralized system currently performs poorer than centralized water from an environmental perspective [9]. Other authors [8] performed a comparative LCA for greywater treatment within a circular economy framework and evaluated the environmental impacts of three greywater treatment alternatives (i.e., photocatalysis, photovoltaic solar-driven photocatalysis, and membrane biological reactor). Their results showed that photovoltaic photocatalysis driven by solar energy is the most sustainable scenario from the environmental point of view [8].

In this scenario, the implementation of a circular economy strategy results in a promising approach [8]. However, LCA studies performed relying on experimental data for the analysis of the environmental impacts of crops irrigated with reclaimed water are still missing [27], although irrigation plays a critical role in boosting crop yield; furthermore, 40% of freshwater global resources are consumed by agricultural production [28].

In particular, to the best of our knowledge, the literature lacks LCA and LCC studies concerning the impact of RWH systems for agricultural purposes, particularly as tool in the decision-making process. In this paper, the authors investigated the environmental and economic impact of a conventional water storage system, as a pond, against an innovative flexible water storage system (FWSS) that could bring about practical advantages to farmers because of its flexibility and the easy-to-move feature. A comparison has been performed via LCA and LCC assessment starting for the hypothesis that 400 m^3 (average commercial pond's volume capacity available on the market) of rainwater can be collected and stored locally for crop irrigation purposes. Furthermore, there are not studies evaluating the ecoefficiency of different rainwater harvesting systems. For this reason, this study fills a knowledge gap in the current literature.

2. Materials and Methods

Farms usually rely on underground reservoir or on aqueduct or, sometime, on channels that naturally occur outside the field during the winter to pump the water needed for watering plants or cleaning machineries. All those sources are temporary available. Thus, it is important to catch as much water as possible during the fall-winter season and store it for the following dry season.

Water storage in ponds is quite common in farms. Thus, the research focused on the economic and environmental sustainability assessment of two systems for water harvesting and storage: the pond and the flexible water storage system (FWSS) (Figure 1).

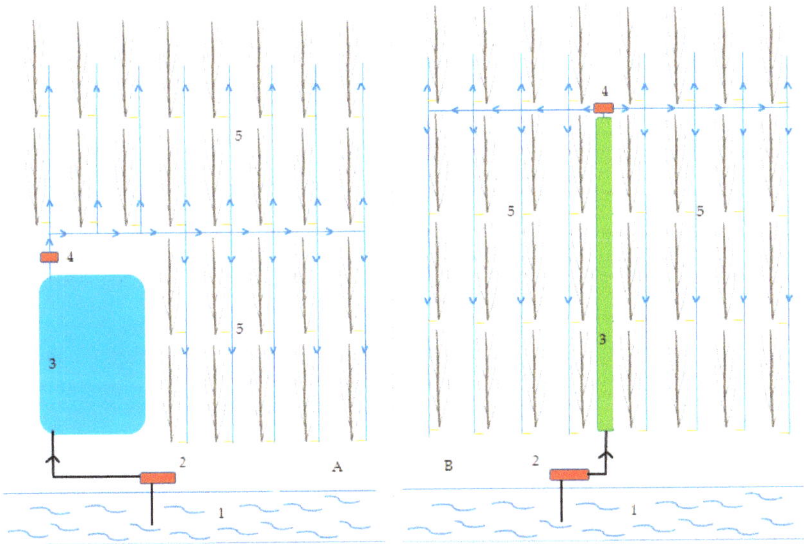

Figure 1. Schematic view of the pond (**A**) and the FWSS (**B**) meant for agroforestry application. Numbers refer to the main components of both systems: (1) seasonal water stream, (2) loading system (including electric pump, pipes and connections), (3) water storage system, (4) electric pump for water delivery, (5) water usage (e.g., irrigation system).

2.1. Pond

Ponds represent a common strategy for water storage because they are relatively easy to build and low demanding in maintenance although they exhibit short lifespan of 5–10 years that could represent a limit for the system [29]. Furthermore, there are other drawbacks that usually are not taken into account like the permanent disturbance of the soil, the reduction of the arable surface which reduces the available arable land and other concerns related to water quality and safety. Moreover, in areas with high evaporation potential ponds are not very suitable [30] and need to be covered with shade net [29].

Building a pond implies permanent changes in the soil, particularly the shallower horizons which are more fertile and suitable for cropping. In order to accommodate the assumed 400 m^3 of water, authors made the hypothesis of removing an equivalent volume of soil (336.6 m^2 and 1.4 m depth) by using a 90-kW excavator. According to the estimations, digging requires 20.5 h and 328.1 l of fuel (data not shown). Consequently, a double layer 1350 g m^{-2} PVC is applied. A detailed list of the components is shown in Table 1.

Table 1. Summary table of components involved in the loading, storing and water distribution in the pond system.

Item	Unit	Quantity	Unitary Weight (kg)		Total Weight (kg)	
			Polyvinyl Chloride	Metal	Polyvinyl Chloride	Metal
Loading system						
Electric water pump	(n)	1.00	-	7.02		7.02
Electric cable	(m)	30.00	0.11	0.07	3.24	2.16
Socket	(n)	1.00	0.12	0.06	0.12	0.06
Pipe connection-equal elbow	(n)	2.00	0.18		0.35	
Pipe connection-adapter	(n)	4.00	0.15		0.60	
Anti-cross flow	(n)	1.00	0.16		0.16	
Ball valve	(n)	2.00	0.18		0.35	
Pipe	(m)	330.00	0.31		102.30	
Pond						
Double PVC layer	(m^2)	477.56	2.35		1122.27	
Fuel	(n)	328.1			328.10	
Irrigation system						
Electric water pump	(n)	1.00		7.02		7.02
Electric cable	(m)	330.00	0.11	0.07	35.64	23.76
Pipe connection-equal elbow	(n)	2.00	0.13		0.26	
Pipe connection-adapter	(n)	4.00	0.11		0.45	
Ball valve	(n)	2.00	0.13		0.26	
Pipe	(n)	400.00	0.23		93.00	
Dripper	(n)	135.00	0.02		2.70	
Total					1689.79	40.01

The Flexible Water Storage System

The flexible water storage system (FWSS) is an alternative solution to ponds. Interestingly, it can be easily folded and moved elsewhere according to the domestic needs of the farm; the installation does not require a concrete base, just a little slope is desired to ease the outflow of the water, which is ensured by a secondary electric pump, though. Contrary to the pond, water is not directly exposed to sunlight; thus microbial activities are not promoted and higher water quality is expected [31].

The Flexible Water Storage System is made of polyvinyl chloride (PVC) 930 g m^{-2} thick and equipped with inlet and outlet pipe connections. FWSSs find many applications in agricultural sector for storing non-potable water, wastewater or sewage water produced by livestock. According to Rigamonti et al. (2019) the service life of a rain water harvesting system based on a polyethylene storage tank is 50 years [32]. Loading is performed by an electric water pump that pumps the water via a filter from a near seasonal water stream directly into the FWSS. When the tank reaches its maximum capacity, the blow-off valve opens preventing over-pressure. The water can be stored as long as it is needed without leak of water or smell. During the dry season the water can be used for irrigation to reduce the exploitation of the underground water. Components of the FWSS are listed in Table 2.

Table 2. Summary table of components involved in the loading, storing and water distribution in the FWSS.

Item	Unit	Quantity	Unitary Weight (kg)		Total Weight (kg)	
			Polyvinyl Chloride	Metal	Polyvinyl Chloride	Metal
Loading system						
Electric water pump	(n)	1.00		7.02		7.02
Electric cable	(m)	30.00	0.11	0.07	3.24	2.16
Socket	(n)	1.00	0.12	0.06	0.12	0.06
Pipe connection-equal elbow	(n)	2.00	0.18		0.35	
Pipe connection-adapter	(n)	4.00	0.15		0.60	
Anti-cross flow	(n)	1.00	0.16		0.16	
Ball valve	(n)	2.00	0.18		0.35	
Pipe	(m)	330.00	0.31		102.30	
FWSS						
Plastic fabric HPVi09	(m^2)	678.84	0.93		631.32	
Pipe connection-reducer	(n)	2.00	0.30		0.60	
Lid	(n)	1.00	0.35		0.35	
Blow-off valve	(n)	1.00	0.15	0.29	0.15	0.29
Ball valve	(n)	4.00		0.80		3.20
Irrigation system						
Electric water pump	(n)	1.00		7.02		7.02
Electric cable	(m)	330.00	0.11	0.07	35.64	23.76
Pipe connection-equal elbow	(n)	2.00	0.13		0.26	
Pipe connection-adapter	(n)	4.00	0.11		0.45	
Ball valve	(n)	2.00	0.13		0.26	
Pipe	(n)	400.00	0.23		93.00	
Dripper	(n)	135.00	0.02		2.70	
Total					871.84	43.50

2.2. LCA and LCC Methods

The environmental impact analysis was carried out using the life cycle assessment methodology (LCA) according to UNI EN ISO 14040:2006 [19] and UNI EN ISO 14044:2006 [20], including the following statements: (a) Goal definition and scoping; (b) life cycle inventory; (c) life cycle impact assessment; (d) life cycle interpretation and improvement. Moreover, the study followed the steps in LCA with the corresponding steps in life cycle costing (LCC) introduced in parallel. Life cycle costing (LCC) is a methodology that aimed to assess the costs across the entire life cycle of a product [33] focusing on the cost at each stage [34].

2.2.1. Boundary of the System for the Life Cycle Assessment (LCA) and Life Cycle Costing (LCC) Analysis

The considered system (for LCA and LCC analysis) is defined by all the processes that occurred during the production and installation phases of two different water tanks (Figure 1). The functional unit is 1 m^3 of storable water of the studied RWH systems. It represents the reference unit used to quantify all inputs and outputs from the studied systems [9].

2.2.2. Life Cycle Inventory Analysis

Primary data of the materials used for the construction of the pond were obtained through interviews to local enterprises, which main activity consists of ponds' construction or, in general, digging jobs. For the FWSS, the primary data was calculated according to [35,36]. Secondary data were obtained by Simapro code database 8.0.2 (Prè Consultants, Amersfoort, The Netherlands) (Tables 3 and 4).

Table 3. Technical data of the pond.

Data	PVC (kg)	Iron (kg)	Diesel (kg)
Loading system	107.11	9.23	
Pond	1122.27		278.89
Irrigation system	132.31	30.77	

Source: data collected from either datasheet or direct weighting of spares.

Table 4. Technical data of the FWSS.

Data	PVC (kg)	Iron (kg)	Diesel (kg)
Loading system	107.11	9.23	
FWSS	632.42	3.49	
Irrigation system	132.31	30.77	

Source: collected from either datasheet or direct weighting of spares.

2.2.3. Life Cycle Impact Assessment

Environmental impacts per m^3 of storable water was assessed using both ReCiPe2008 [37] and GWP100 methods. In particular, ReCiPe 2008 method includes categories of environmental impact and environmental damage, i.e., at the midpoint and endpoint level. Initially, the inventory data were associated to the midpoint level using factors of characterizations. Lately, they have been converted and clustered into the endpoint level considering three damage categories (i.e., HH, EC, RE damage categories) and by using weighting factors. It is important to underline that it was applied the method ReCipe Endpoint (H)/ Europe ReCipe H/A by considering weighing factors referred to the mean values of the hierarchical perspective. Moreover, GHG emission was chosen to link the environmental issue to the economic aspect of the water harvesting systems in order to determine their eco-efficiency [38]. The carbon footprint was defined as the sum of all GHGs emitted within the system boundary and expressed in CO_2 equivalent according to IPCC 2007 method (100-year life span). A parallel economic assessment is integrated with LCA using a life cycle perspective. It is important to underline that the economic sustainability is an important aspect to consider for farmers.

2.2.4. Economic Assessment

The possibility of conducting a LCA study integrated with Life cycle costing (LCC) contributes to improve the ecoefficiency of farms [39], and thus reducing their impacts on the environment, while reducing costs [40].

A conventional cradle-to-gate LCC was applied here encompassing the assessment of all costs associated with the life cycle of both RWH systems studied.

The cost of the items included in the analysis referred to raw materials, services, other operating expenses, and labor costs. The economic data (Tables 5 and 6) derive from informal local market as proposed in other studies [41,42].

Table 5. Economic data of the pond.

Data	Total Costs (Euro)
Loading system	669.47
Pond	6356.93
Irrigation system	1140.37
Total	**8166.77**

Total costs include raw materials, services, other operating expenses, and labor costs for each step. Source: data retrieved from informal local market.

Table 6. Economic data of the FWSS.

Data	Total Costs (Euro)
Loading system	669.47
FWSS	4968.00
Irrigation system	1140.37
Total	**6777.84**

Total costs include raw materials, services, other operating expenses, and labor costs for each step. Source: data retrieved from informal local market.

3. Results and Discussion

A large majority of literature deals with the environmental impact of water management in agriculture purely in terms of water and energy used [23–26].

It is clear that irrigation might tip the scale towards a less sustainable scenario as observed by Stephenson et al. (2010) [23]. Some authors have evaluated the emissions generated by water harvesting systems and the related costs [32], even if the majority of the studies focused on the impact of RWH systems in urban environment [18,43–46].

In fact, rain and storm water harvesting systems are widely used in urban and agricultural areas especially where the weather conditions are unfavorable, with periods of drought alternating with periods characterized by floods and torrential rains. These aspects were already highlighted by Ghimire et al. (2017) that among the benefits of RWH indicated the reduction of stormwater runoff and combined sewers overflows events, as well as the potential impact reduction on the environment and human health, remarking a lack of understanding in the magnitude of these positive effects [13]. This is especially true in agriculture where, to the best of our knowledge, few studies analyzed the environmental and economic impacts of RWH infrastructure.

3.1. Environmental Assessment

The impact analysis allowed to identify the RWH infrastructure and its installation which has higher environmental impact. It is important to underline that with the weighing it is possible to assess the importance of each category of impact obtaining aggregate results as damage categories [47]; while the characterization permits to quantify the general impacts concerning different impacts categories [47].

Figure 2 reports the LCA results of each studied RWH system at endpoint level. Comparing the outcomes of weighing (Figure 2) and characterization (Figure 3) helps to identify the environmental performance and impacts of each RWH system. The highest damage categories were resources and human health, while ecosystem damage was the lowest in all systems.

The pond causes the highest impact on resources and human health due to used raw material. Polyvinyl chloride (PVC) production had the highest impact on all damage categories (especially in the pond system); this finding was due to the characteristics of PVC production. Additionally, Ghimire et al. (2017) found out that storage tanks represent the second most important cause of environmental impact (after energy usage) [13]. In order to reduce the impacts, the material used for the construction of a RWH device represents an important aspect to take in consideration. This is true also for the life span and volume of the storage tank that may lead to differing impacts [48]. In fact, Ghimire at al. (2014) observed lower impact of polyethylene (PE) when compared with a concrete storage tank, even if the latter has an expected life span of 70 years (50 years for the PE storage tank) [48]. According to Ghimire et al. (2017), the PE storage tank resulted as a good alternative to the RWH fiberglass storage tank that was less sustainable for the ozone depletion and freshwater withdrawal impact categories [13]. However, to the best of our knowledge, no other studies reported information about the specific impact generated by PVC used for RWH construction.

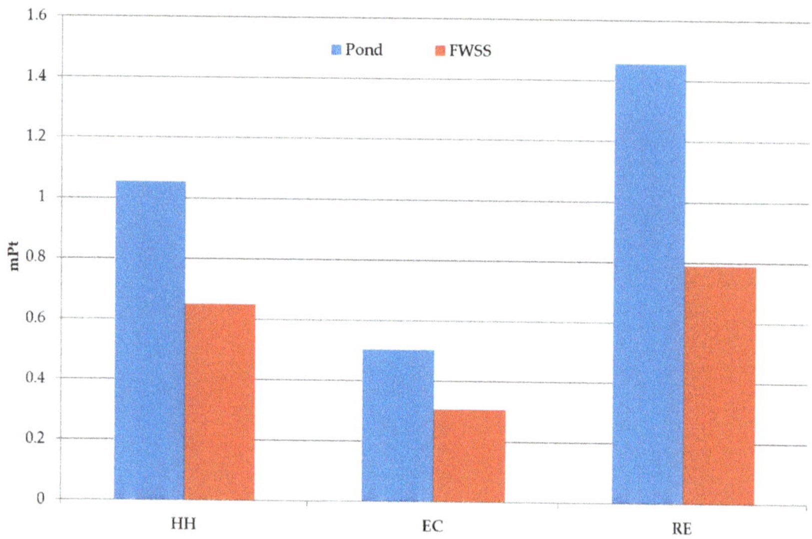

Figure 2. Result of the weighing, comparison of two different RWH systems with functional unit of 1 m^3 of storable water. The acronyms of the different damage categories are HH (Human Health), EC (Ecosystem), RE (Resources).

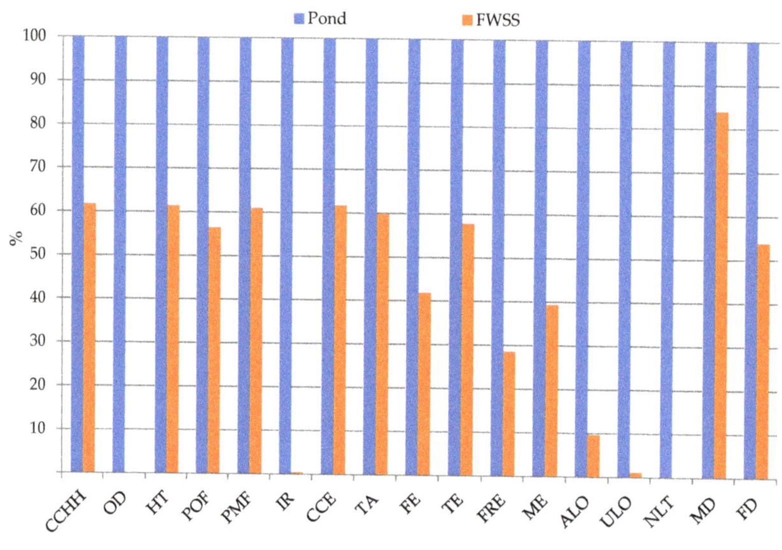

Figure 3. Result of the characterization, comparison of two different runoff water storage systems with functional unit of 1 m^3 of storable water. The values are expressed as percentages in relation to the 100% given to RWH systems with the biggest impact in each category (i.e., Pond = 100% in all the considered impact categories). The acronyms of the different impact categories are CCHH (climate change human health), OD (ozone depletion), HT (human toxicity), POF (photochemical oxidant formation), PMF (particulate matter formation), IR (ionizing radiation), CCE (climate change ecosystems), TA (terrestrial acidification), FE (freshwater eutrophication), TE (terrestrial ecotoxicity), FRE (freshwater ecotoxicity), ME (marine ecotoxicity), ALO (agricultural land occupation), ULO (urban land occupation), NLT (natural land transformation), MD (metal depletion), FD (fossil depletion).

The characterization analysis (Figure 3) showed the environmental performance of each RWH system in relative terms reporting midpoint environmental impacts.

The analysis showed that the pond system is more impactful than the FWSS in all impact categories. In particular, PVC production was the most impactful phase (11 out of 17 impact categories) regardless of the RWH system; this was due to the characteristics of the PVC production itself. In particular, for each studied system, the environmental impacts on CCHH category were due to carbon dioxide emissions coming from PVC production. This production also impacts the HT category (due to dioxin emission), POF (caused by nitrogen oxides emissions), PMF (due to sulfur dioxide emissions), CCE (caused by carbon dioxide emissions) and TA (due to sulfur dioxide emissions) impact categories. Moreover, the PVC production impacts other categories: FE (caused by phosphorus emissions in the water) and TE (due to chlorine emissions) impact categories. In addition, the impact of PVC production on the FRE category was due to nickel emission in the water for the pond system and copper emission (in the water) for the FWSS. The impacts on the ME category were due to chlorine emissions coming from PVC production. Finally, the PVC production also impacted the fossil depletion (FD) category, and this was because of the energy necessary for the production.

It is interesting to observe that 3 out of 17 impact categories (i.e., OD, IR and NLT impact categories) are affected only by environmental impact due to the pond system, while these categories are not involved in the FWSS. In fact, the extraction of oil and the production of fuel and its combustion during the excavation contributed to OD, IR and NLT impact categories only for the pond system.

Moreover, it is important to underline that the pond system is more impactful than the FWSS in the ALO and ULO impact categories due to a different occupation of land between the two water storage systems. In fact, for the same volume of stored water, unlike the pond, the FWSS does not permanently occupy arable land and can be installed on the uncultivated soil of the farm (e.g., over ditches or between rows of permanent crops that do not require machine passage). This is a major advantage especially in farms where space is a limiting factor.

Finally, the impacts on the MD category were due to the irony parts of excavator (used to create the pond), loading and irrigation systems (for both RWH systems).

3.2. The Economic Aspects and Ecoefficiency Analysis

The LCC analysis of the water systems was carried out in similar phases corresponding to LCA standard. Table 7 shows the total costs referred to the functional unit of 1 m^3 of storable water.

Table 7. The economic aspect of each system per 1 m^3 of water.

Data	Total Costs (€ per m^3)
Pond system	20.41
FWSS	16.94

All data are referred to 1 m^3. Source: Authors' elaboration.

Findings showed that the FWSS's costs are slightly lower than the pond system's; in fact, the FWSS shows total costs of 16.94 € per m^3 of storable water, versus 20.41 € per m^3 in the pond system. The current literature lacks studies dealing with costs of an FWSS, though this is quite obvious considering that this system represents an innovation in the sector, which has not been studied yet. On the other hand, several studies analyzed economic aspects related to ponds' construction for rainwater harvesting, with particular reference to irrigation purposes. A literature review carried out by Lasage et al. (2015) reported an average cost for water storage ponds, with dimensions ranging from 30 to 300 m^3, of 17.16 € per m^3 [49]. This value is slightly lower than what was found in the present study. Surprisingly, this higher cost is not related to the different lining material (cement-wire in the cases studied in the review and plastic film in this paper's), considering that cost for lining with cement and wire is generally higher than the cost for the same operation performed with plastic film. Several further studies were carried out especially in India. Reported ponds' construction costs were substantially lower than what was found in the present study. In details, Deshmukh et al. (2016) reported 1.61 € per m^3 [50], while 3.71 € per m^3 were found for a pond lined through an HDPE (high-density polyethylene) geomembrane by Reddy at al., (2020) [51]. Finally, Shalander et al. (2016) reported construction costs for unlined ponds in India ranging from 1.80 to 4.35 € per m^3 [52]. Such markedly lower construction costs for ponds are related to both the absence of lining material (in most cases) and to the lower labor costs compared to the Italian ones. Interestingly, Shalander et al. (2016) also reported construction costs for a particular rainwater storage system typical of the Jodhpur region, locally named Tanka [52]. This consists of an underground cistern made of concrete. This system is, under some aspects, comparable to the FWSS, considering that water is not stored in direct contact with air. Construction costs of a Tanka resulted equal at 21.25 € per m^3—therefore higher than FWSS ones, highlighting how such innovative water storage system is also suitable outside of Europeon contexts.

An important aspect of the ecoefficiency analysis of two different water storage systems was the variability of their main components, as was demonstrated by differences in carbon footprint (as an environmental indicator) and life cycle costs (as an economic indicator) in relation to RWH systems.

The highest value of the carbon footprint and costs was obtained for the pond system, while the FWSS exhibited the lowest value of GHG emission and aggregated costs (Figure 4). These findings showed that the FWSS was the best solution under the economic and environmental point of view. In fact, the FWSS and the pond system cost 16.94 € with an emission of 17.4 gr CO_2 eq per m^3 of storable water and 20.41 € with an emission of 28.2 gr CO_2 eq per m^3 of storable water, respectively.

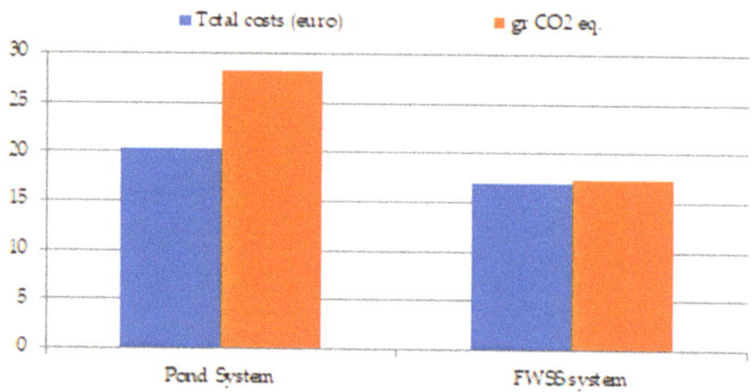

Figure 4. Combined results of the LCC and GWP (LCA analysis) of each RWH system per m^3 of storable water. Source: Authors' elaboration.

4. Conclusions

The present work aimed to perform the environmental and economic analysis of two different RWH systems for irrigation purposes. In particular, LCA and LCC assessments were carried out to compare, under environmental and economic aspect, the water storage of a typical pond and an innovative flexible water storage system (FWSS). The current literature encompasses only few studies assessing the environmental and economic performance of different rainwater harvesting strategies in agriculture. For this reason, the present work represents a first attempt of such kind of evaluation in the topic—mainly regarding the evaluation of eco-efficiency—and the first step to fulfill such knowledge gap.

Findings showed that the FWSS performed better in both environmental and economic aspects, resulting in a suitable and sustainable solution for water harvesting in agriculture.

Evaluating the environmental and economic performance of a given system and carrying out comparisons among possible alternatives is crucial for a proper decision-making process. Considering the importance of water scarcity and water harvesting topics in a circular economy framework, carrying out this kind of scientific analysis could provide an important contribution to the decision-makers. In particular, our findings are useful not only for academics but also for farmers and practitioners working in the topic of water management.

Further studies should focus on a real case study, providing tangible evaluation of the environmental and economic performance based on existing RWH systems. The sustainability of such system highlighted by our results—along with the valuable features of the FWSS associated with flexibility and nonpermanent disturbance of the environment—paves the way for interesting abroad applications, particularly where water scarcity jeopardizes human health and food security in arid countries of the globe.

Author Contributions: Conceptualization N.P., A.S., W.S. and F.L.; formal analysis, N.P.; investigation, N.P., W.S. and F.L.; data curation and methodology N.P.; writing—original draft preparation N.P., A.S., W.S. and F.L.; In particular, Introduction paragraph, N.P and A.S.; Materials and Methods paragraphs N.P. and W.S.; Results and Discussion paragraph N.P., A.S. and F.L; Conclusions paragraph N.P., A.S., W.S. and F.L.; writing—review and editing N.P., A.S., W.S. and F.L.; supervision, project administration and funding acquisition, L.P. All authors have read and agreed to the published version of the manuscript.

Funding: This research was funded by the AGROENER project (D.D. n. 26329, 1 April 2016) funded by the Italian Ministry of Agriculture (MiPAAF) and the APC was funded by AGROENER project. The ideas expressed do not represent those of the Italian Ministry of Agriculture (MiPAAF).

Conflicts of Interest: The authors declare no conflict of interest.

References

1. European Environment Agency. *Water Scarcity and Drought in the European Union*; Rep. KH-30-09-180-EN-D; European Environment Agency: København, Denmark, 2010; p. 6.
2. Benabdelouahab, T.; Gadouali, F.; Boudhar, A.; Lebrini, Y.; Hadria, R.; Salhi, A. Analysis and trends of rainfall amounts and extreme events in the Western Mediterranean region. *Theor. Appl. Climatol.* **2020**, *141*, 309–320. [CrossRef]
3. Changming, L.; Jingjie, Y.; Kendy, E. Groundwater Exploitation and Its Impact on the Environment in the North China Plain. *Water Int.* **2001**, *26*, 265–272. [CrossRef]
4. Gomiero, T.; Pimentel, D.; Paoletti, M.G. Environmental impact of different agricultural management practices: Conventional vs. organic agriculture. *CRC. Crit. Rev. Plant Sci.* **2011**, *30*, 95–124. [CrossRef]
5. Helmreich, B.; Horn, H. Opportunities in rainwater harvesting. *Desalination* **2009**, *248*, 118–124. [CrossRef]
6. Uribe, I.O.; Mosquera-Corral, A.; Rodicio, J.L.; Esplugas, S. Advanced technologies for water treatment and reuse. *AIChE* **2015**, *61*, 3146–3158. [CrossRef]
7. Wilcox, J.; Nasiri, F.; Bell, S.; Rahaman, M.S. Urban water reuse: A triple bottom line assessment framework and review. *Sustain. Cities Soc.* **2016**, *27*, 448–456. [CrossRef]
8. Dominguez, S.; Laso, J.; Margallo, M.; Aldaco, R.; Rivero, M.J.; Irabien, Á.; Ortiz, I. LCA of greywater management within a water circular economy restorative thinking framework. *Sci. Total Environ.* **2018**, *621*, 1047–1056. [CrossRef]
9. Yan, X.; Ward, S.; Butler, D.; Daly, B. Performance assessment and life cycle analysis of potable water production from harvested rainwater by a decentralized system. *J. Clean. Prod.* **2018**, *172*, 2167–2173. [CrossRef]
10. Krishna, R.S.; Mishra, J.; Ighalo, J.O. Rising Demand for Rain Water Harvesting System in the World: A Case Study of Joda Town, India. *World Sci. News* **2020**, *146*, 47–59.
11. Karim, M.R.; Bashar, M.Z.I.; Imteaz, M.A. Reliability and economic analysis of urban rainwater harvesting in a megacity in Bangladesh. *Resour. Conserv. Recycl.* **2015**, *104*, 61–67. [CrossRef]
12. Jiang, Z.-y.; Li, X.-y.; Ma, Y.-j. Water and Energy Conservation of Rainwater Harvesting System in the Loess Plateau of China. *J. Integr. Agric.* **2013**, *12*, 1389–1395. [CrossRef]
13. Ghimire, S.R.; Johnston, J.M.; Ingwersen, W.W.; Sojka, S. Life cycle assessment of a commercial rainwater harvesting system compared with a municipal water supply system. *J. Clean. Prod.* **2017**, *151*, 74–86. [CrossRef]
14. Qi, Q.; Marwa, J.; Mwamila, T.B.; Gwenzi, W.; Noubactep, C. Making rainwater harvesting a key solution for water management: The universality of the Kilimanjaro Concept. *Sustainability* **2019**, *11*, 5606. [CrossRef]

15. Hussain, F.; Hussain, R.; Wu, R.-S.; Abbas, T. Rainwater harvesting potential and utilization for artificial recharge of groundwater using recharge wells. *Processes* **2019**, *7*, 623. [CrossRef]
16. Kisakye, V.; Van der Bruggen, B. The viability of artificial surface treatments as a mechanism for domestic rain water harvesting. *Phys. Chem. Earth Parts A/B/C* **2018**, *107*, 8–18. [CrossRef]
17. Vialle, C.; Busset, G.; Tanfin, L.; Montrejaud-Vignoles, M.; Huau, M.C.; Sablayrolles, C. Environmental analysis of a domestic rainwater harvesting system: A case study in France. *Resour. Conserv. Recycl.* **2015**, *102*, 178–184. [CrossRef]
18. Morales-Pinzón, T.; Lurueña, R.; Rieradevall, J.; Gasol, C.M.; Gabarrell, X. Financial feasibility and environmental analysis of potential rainwater harvesting systems: A case study in Spain. *Resour. Conserv. Recycl.* **2012**, *69*, 130–140. [CrossRef]
19. ISO (International Organization for Standardization). *14040:2006—Environmental Management—Life Cycle Assessment—Principles and Framework*; ISO: Geneva, Switzerland, 2006.
20. ISO (International Organization for Standardization). *14044:2006—Environmental Management—Life Cycle Assessment—Requirements and Guidelines*; ISO: Geneva, Switzerland, 2006.
21. ISO (International Organization for Standardization). *15686-5:2017—Buildings and Constructed Assets—Service Life Planning—Part 5: Life-Cycle Costing*; ISO: Geneva, Switzerland, 2017.
22. Gluch, P.; Baumann, H. The life cycle costing (LCC) approach: A conceptual discussion of its usefulness for environmental decision-making. *Build. Environ.* **2004**, *39*, 571–580. [CrossRef]
23. Stephenson, A.L.; Von Blottnitz, H.; Brent, A.C.; Dennis, J.S.; Scott, S.A. Global warming potential and fossil-energy requirements of biodiesel production scenarios in South Africa. *Energy Fuels* **2010**, *24*, 2489–2499. [CrossRef]
24. Mohamad, R.S.; Verrastro, V.; Cardone, G.; Bteich, M.R.; Favia, M.; Moretti, M.; Roma, R. Optimization of organic and conventional olive agricultural practices from a Life Cycle Assessment and Life Cycle Costing perspectives. *J. Clean. Prod.* **2014**, *70*, 78–89. [CrossRef]
25. Buratti, C.; Fantozzi, F. Life cycle assessment of biomass production: Development of a methodology to improve the environmental indicators and testing with fiber sorghum energy crop. *Biomass Bioenergy* **2010**, *34*, 1513–1522. [CrossRef]
26. Figueiredo, F.; Castanheira, E.G.; Freire, F. Life-cycle assessment of irrigated and rainfed sunflower addressing uncertainty and land use change scenarios. *J. Clean. Prod.* **2017**, *140*, 436–444. [CrossRef]
27. Romeiko, X.X. A comparative life cycle assessment of crop systems irrigated with the groundwater and reclaimedwater in Northern China. *Sustainability* **2019**, *11*, 2743. [CrossRef]
28. Eshel, G.; Martin, P.A. Diet, energy, and global warming. *Earth Interact.* **2006**, *10*, 1–17. [CrossRef]
29. WFP. Water Harvesting in Practice: Towards Building Resilient Livelihoods in Semi-Arid Zones. In *Field Practitioners Guide No. 2. Rural Resilience*; World Food Programme: Nairobi, Kenya, 2018.
30. Philp, M.; Mcmahon, J.; Heyenga, S.; Marinoni, O.; Jenkins, G.; Maheepala, S.; Greenway, M. *Review of Stormwater Harvesting Practices*; Urban Water Security Research Alliance Technical Report No. 9; The Urban Water Security Research Alliance: City East, QLD, Australia, 2008. Available online: https://publications.csiro.au/rpr/download?pid=procite:6057d2b6-42b7-4d03-9fd7-ddd451b9d269&dsid=DS1 (accessed on 4 March 2021).
31. Avnimelech, Y. Control of microbial activity in aquaculture systems: Active suspension ponds. *World Aquac. Rouge* **2003**, *34*, 19–21.
32. Ghimire, S.R.; Johnston, J.M. Holistic impact assessment and cost savings of rainwater harvesting at the watershed scale. *Elementa* **2017**, *5*, 9. [CrossRef] [PubMed]
33. Rigamonti, L.; Borghi, G.; Martignon, G.; Grosso, M. Life cycle costing of energy recovery from solid recovered fuel produced in MBT plants in Italy. *Waste Manag.* **2019**, *99*, 154–162. [CrossRef]
34. Brandão, M.; Clift, R.; Basson, L. A life-cycle approach to characterising environmental and economic impacts of multifunctional land-use systems: An integrated assessment in the UK. *Sustainability* **2010**, *2*, 3747–3776. [CrossRef]
35. Labaronne-Citaf Website. Available online: labaronne-citaf.com (accessed on 10 January 2021).
36. Leffio, D. Personal comunication. 2021.
37. Goedkoop, M.; Heijungs, R.; Huijbregts, M.; De Schryver, A.; Struijs, J.; Van Zelm, R. *ReCiPe 2008. A Life Cycle Impact Assessment Method which Comprises Harmonised Category Indicators at the Midpoint and the Endpoint Level*; Ministerie van VROM Rijnstraat: Den Haag, The Netherlands, 2009; Volume 1, pp. 1–126.
38. Palmieri, N.; Suardi, A.; Alfano, V.; Pari, L. Circular economy model: Insights from a case study in south Italy. *Sustainability* **2020**, *12*, 3466. [CrossRef]
39. Pari, L.; Suardi, A.; Stefanoni, W.; Latterini, F.; Palmieri, N. Environmental and economic assessment of castor oil supply chain: A case study. *Sustainability* **2020**, *12*, 6339. [CrossRef]
40. Baum, R.; Bieńkowski, J. Eco-efficiency in measuring the sustainable production of agricultural crops. *Sustainability* **2020**, *12*, 1418. [CrossRef]
41. Palmieri, N.; Suardi, A.; Latterini, F.; Pari, L. The eucalyptus firewood: Understanding consumers' behaviour and motivations. *Agriculture* **2020**, *10*, 512. [CrossRef]
42. Palmieri, N.; Suardi, A.; Pari, L. Italian consumers' willingness to pay for eucalyptus firewood. *Sustainability* **2020**, *12*, 2629. [CrossRef]
43. Moore, T.L.C.; Hunt, W.F. Predicting the carbon footprint of urban stormwater infrastructure. *Ecol. Eng.* **2013**, *58*, 44–51. [CrossRef]

44. Ghimire, S.R.; Johnston, J.M.; Garland, J.; Edelen, A.; Ma, X.C.; Jahne, M. Life cycle assessment of a rainwater harvesting system compared with an AC condensate harvesting system. *Resour. Conserv. Recycl.* **2019**, *146*, 536–548. [CrossRef] [PubMed]
45. Morales-Pinzón, T.; Rieradevall, J.; Gasol, C.M.; Gabarrell, X. Modelling for economic cost and environmental analysis of rainwater harvesting systems. *J. Clean. Prod.* **2015**, *87*, 613–626. [CrossRef]
46. Devkota, J.; Schlachter, H.; Apul, D. Life cycle based evaluation of harvested rainwater use in toilets and for irrigation. *J. Clean. Prod.* **2015**, *95*, 311–321. [CrossRef]
47. Arrivas Bajardi, C.; Fiore, M.; Breedveld, L.; Giaimo, L.; Notaro, A. *Certifcazione Ambientale di Prodotti Agroalimentari LCA Dell'olio D'oliva*; 2009. Available online: http://www.to-be.it/wp-content/uploads/2015/07/LCA-dellolio-doliva.pdf (accessed on 4 March 2021).
48. Ghimire, S.R.; Johnston, J.M.; Ingwersen, W.W.; Hawkins, T.R. Life Cycle Assessment of Domestic and Agricultural Rainwater Harvesting Systems. *Environ. Sci. Technol.* **2014**, *48*, 1–27. [CrossRef]
49. Lasage, R.; Verburg, P.H. Evaluation of small scale water harvesting techniques for semi-arid environments. *J. Arid Environ.* **2015**, *118*, 48–57. [CrossRef]
50. Deshmukh, G.; Hardaha, M.K.; Ahirwar, S.K. A case study on rain water harvesting technologies for tribal area of Madhya Pradesh, India. *Plant Arch.* **2016**, *16*, 151–156.
51. Reddy, K.S.; Ricart, S.; Maruthi, V.; Pankaj, P.K.; Krishna, T.S.; Reddy, A.A. Economic Assessment of Water Harvesting Plus Supplemental Irrigation for Improving Water Productivity of a Pulse–Cotton Based Integrated Farming System in Telangana, India. *Irrig. Drain.* **2020**, *69*, 25–37. [CrossRef]
52. Kumar, S.; Ramilan, T.; Ramarao, C.A.; Rao, C.S.; Whitbread, A. Farm level rainwater harvesting across different agro climatic regions of India: Assessing performance and its determinants. *Agric. Water Manag.* **2016**, *176*, 55–66. [CrossRef]

sustainability

Article

Factors Influencing Consumers' Attitude Towards Biopreservatives

Maria Angela Perito [1,2,*], Emilio Chiodo [1], Annalisa Serio [1], Antonello Paparella [1] and Andrea Fantini [1]

[1] Faculty of Bioscience and Technology for Food, Agriculture and Environment, University of Teramo, 64100 Teramo, Italy; echiodo@unite.it (E.C.); aserio@unite.it (A.S.); apaparella@unite.it (A.P.); afantini@unite.it (A.F.)

[2] UR ALISS, INRAE, Université Paris-Saclay, 94205 Ivry-sur-Seine, France

* Correspondence: maperito@unite.it

Received: 5 November 2020; Accepted: 8 December 2020; Published: 10 December 2020

Abstract: Biopreservatives have received considerable attention in recent years as natural alternatives to synthetic preservatives. This seems to be a response to an increased demand for natural and organic foods. This study investigates the potential market for products enriched with biopreservatives in Italy. Data were collected from a sample of Italian consumers (N = 479) using a web-based survey. The main results indicate that 64% of respondents declared themselves to be willing to consume biopreservatives only if they replaced synthetic preservatives. Principal component analysis (PCA) was applied to reduce the number of variables. The factorial scores of the components obtained from PCA were used for a Cluster Analysis related to consumers' perceptions about biopreservatives. Moreover, the survey highlights that the respondents had positive opinions about biopreservatives, although they showed difficulty in perceiving the exact meaning of the term. The study could provide useful implications for food manufacturers and facilitate the design of marketing strategies for foods enriched with biopreservatives.

Keywords: biopreservatives; shelf life; essential oil; organic foods; consumers' attitude; willingness to pay

1. Introduction

Delivering food in good conditions from the production site to the consumer often requires the use of additives. Food additives are defined by Regulation (EC) No 1333/2008 [1] as "substances that are not normally consumed as food itself but are added to food intentionally for a technological purpose", to promote food safety and extend product shelf life. In fact, traditionally, the control of food-spoiling and pathogenic microorganisms is ensured by the use of synthetic substances. Despite the benefits that some food additives apparently have, consumers feel that additives should be reduced in their foods and that they are 'bad' for their health [2]. Arbindra et al. [3] and Shim et al. [4] pointed out that the consumers' perception of food additives is generally negative.

Consumers face many food choices associated with food additives every day. In general, they prefer food with no additives, but if not available, consumers will choose foods containing natural additives over synthetic analogues [5–11]. In particular, natural additives [12] have been gaining interest from producers and consumers. Generally, consumers have a high-risk perception of industrially produced and processed foods [13]. In contrast, the word "natural" on food labels evokes mainly positive associations [14]. According to Lockie et al. [13], Rozin [14] and Siipi [15] "naturalness" is an attribute that enhances the positive perception of foods, making these products more desirable to the corresponding non-natural ones. During past decades, the market demand for natural and organic products has grown in many industrialized countries [16–18], as well as the request for ready-to-eat food products and fresh fruits and vegetables [19]. Consumers consider ready-to-eat products as very perishable and perceive freshness as the most important factor influencing food choice, both during purchase and consumption [20–22]. Nevertheless, these products are the most exposed to contamination by spoiling and even pathogenic microorganisms, thus requiring particular care to ensure consumer safety and product shelf life [23].

In this general trend, consumer attention is growing towards the properties of medicinal plants (especially organic ones) and the production of essential oils (EOs) and hydrosols (or hydrolates), which are both obtained during the distillation of aromatic plants. The use of EOs as natural preservatives may be useful to improve food safety, mainly because of their antioxidant and antimicrobial activities [24]. EOs are hydrophobic compounds and can be directly applied to food products by means of surface treatments (application on the surface by dipping, spraying, brushing, and panning), or included within the matrix (e.g., in minced meat). EOs can be also enclosed in antimicrobial edible films or coatings, generally made of biomaterials such as proteins or carbohydrates, with the addition of plasticizers and, eventually, jellying agents [25]. The micro- or nano-encapsulation of EOs and bioactive molecules into edible carriers (e.g., cyclodextrins, proteins, etc.) is another technique that allows us to increase the stability and bioactivity of the active compounds and to reduce the impact on the flavour of the food product. On the contrary, hydrosols are hydrophilic, and therefore they are usually applied to food products by means of washing treatments (e.g., vegetables, fruit, ready-to-eat salads, etc.) [23].

Nevertheless, in spite of these wide possibilities of application, few of the preservatives containing EOs for use in food products are currently available on the market, and they are typically used in the food industry as flavouring agents [26]. Biopreservatives can represent a source of natural alternatives to conventional preservatives to improve food shelf life and safety [27].

The regulations are strengthening to reduce the food-related risk of consumption and to preserve the health of consumers [28], and preservatives are included in the legal category of food additives, according to Regulation (EC) 1333/2008 [1]. However, EOs used as biopreservatives, as long as they are not listed in the EU Regulation on food additives, are usually included in the ingredient list as an ingredient: e.g., organic essential oil of oranges. If the EO is manufactured and distributed according to Regulation (EC) 1334/2008 [29] on food flavourings, Regulation (EU) 1169/2011 [30] on food labelling gives us another option for the ingredient list, which is "natural flavourings". Finally, if the manufacturer can demonstrate that the EO is added to obtain a certain technological purpose during processing, leaving residues that do not present any health risk and do not have any technological effect on the finished product, then it can be considered as a "processing aid" (Regulation (EC) 1333/2008, art. 3), and according to Regulation (EU) 1169/2011, is not reported in the ingredient list.

In any case, among the information contained on the label, the consumers pay particular attention on the expiry date and they consider it an important quality attribute [31,32]. Stranieri and Baldi [33] investigated a sample of Italian consumers, and they highlighted how consumers pay attention to product shelf life, especially for fresh-cut vegetables. As a consequence, considering that these vegetables are very perishable products, the consumers' choices in purchase are mostly guided by the perceived level of freshness [34,35]. In this respect, EOs can have a significant antimicrobial impact. Nevertheless, their use in foods is quite limited due to both a high cost and a possible adverse impact on sensory characteristics and product acceptability [36]. Moreover, from a regulatory point of view, there is not much on the topic of natural preservatives, and in fact the term "biopreservative" is neither regulated nor used on the label.

Therefore, the main objective of this research was to identify the potential for the development of the biopreservatives market through the analysis of consumer perception and acceptance of natural food preservatives as an alternative to synthetic preservatives. In particular, to understand the real consumer acceptance of biopreservatives, we concentrated our analysis on three fresh products (fresh-cut vegetables, meats and cheeses). The study also aims to provide information for food manufacturers regarding the most appropriate marketing levers for the enhancement of biopreservatives.

2. Materials and Methods

The direct survey of consumers was concerned with the perception of natural preservatives as an element for both improving the shelf life of food products and replacing synthetic preservatives with natural preservatives, with the aim of understanding the type of message communicated to the consumer.

The questionnaire was distributed throughout Italy by ADICONSUM (Consumers and Environment Defence Association), who administered the questionnaire via email to its members and through Facebook, collecting 479 complete answers. Participation took place in an absolutely anonymous form, and the participants were informed in advance that the data collected would have been treated in an extremely confidential manner and used only for scientific research purposes. However, respondents were recruited from diverse community centres.

The self-administered questionnaire contained questions with closed-ended response alternatives on a five-point Likert-type scale. The content validity of the questionnaire was ascertained by a pre-test to collect elements to assess completeness and clarity of the questionnaire. Specifically, the survey allowed us to investigate the consumer perception and knowledge about biopreservatives and to assess the interest in the purchase and the willingness to pay towards food products treated with natural preservatives, as an alternative to synthetic preservatives.

The questionnaire consists of questions on the knowledge of biopreservatives to investigate the awareness about the preservatives. Some of these questions contained definitions from the EU Regulation, which indirectly allowed us to evaluate whether the legislation is understandable for the consumer or if there is an actual lack of clarity from the regulatory point of view, as highlighted on more than one occasion in the literature about novel food products or technologies [37–39]. In particular, the focus of the investigation was on consumer knowledge of the definition of preservative and the difference between natural and synthetic preservatives. The perception of information that can be transmitted by product labelling was analysed considering both the mandatory and hypothetically voluntary information on the labels [40]. The aim of these questions was to analyse the importance to the consumers of information contained on the label and their perception of different claims [41,42]. Other questions were related to consumer preferences, their willingness to pay for natural preservatives (as percentage of synthetic preservatives), and the use on several food products.

To understand the consumers' acceptance regarding natural preservatives, we also asked questions about their purchasing habits of food products and the frequency of purchase of organic products to verify whether or not there was a correlation between the consumption of organic products and interest in purchasing and willingness to pay for products treated with natural preservatives. According to Dickson-Spillmann et al. [43], consumers consider organic foods to be healthier, uncontaminated, and purer than conventional foods, and not altered or polluted by synthetic additives or by excessive human interference. Finally, we also asked about information relating to personal data (gender, age, profession, level of education, etc.).

A Principal Component Analysis (PCA) was applied (computed using IBM SPSS Statistics-version 20.0.0) on the original data to reduce the initial diversity of a certain number of variables [44]. The factorial scores of the components obtained from PCA were used for a Cluster Analysis related to consumers' perceptions about biopreservatives.

3. Results and Discussion

A total of 479 respondents completed the questionnaire. In the sample, 64% of respondents were 25–44 years old and 58.2% were male. The sample contained a high percentage of graduates. The apparent imbalance of the sample can be considered a strength of the questionnaire, because it allowed us to investigate the capacity of highly educated people to evaluate natural preservatives.

To evaluate the knowledge of consumers, we asked them to define preservatives and natural preservatives. In this respect, 94.8% of the sample answered the question "what is a preservative" correctly, according to the definitions given by Regulation (EC) 1333/2008. As for the term "biopreservative", although 46% of the sample answered correctly, it is true that there is still much uncertainty about the meaning of this term. Indeed, when consumers were asked the difference between natural and synthetic preservatives, only 38.8% of the respondents answered correctly. With regards to the definition of flavourings, 65.8% of the sample answered correctly.

Although most respondents read food labels (87% from often to always), there was no significant correlation between the frequency of reading labels and the knowledge about biopreservatives. Moreover, considering different information that consumers can find on the label, the data obtained demonstrate a greater concern of the respondents for the expiry date and the presence of food additives (Figure 1).

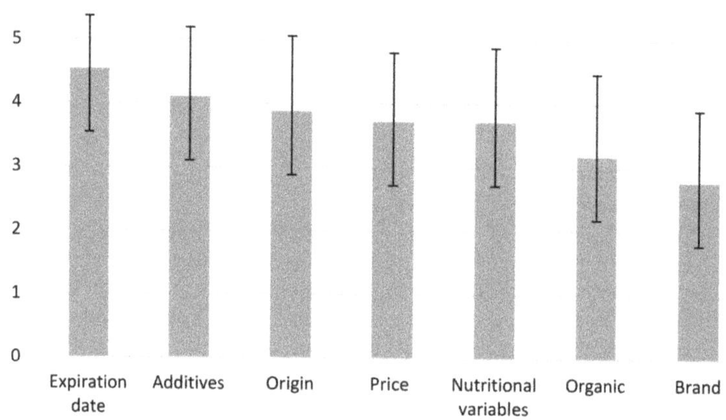

Figure 1. The level of attention paid to different information on the label (mean and standard deviation). (Attributes importance (scale 1 to 5: where 1 = Not important and 5 = Very important)).

Furthermore, the claims "without preservatives", "without added preservatives" and "organic", which are actually used in food labels, have been compared with the claims "natural" or "bio" preservatives, which are not currently included in the labelling regulation, so that we can consider these as hypothetical labels. The results showed that the new claims would be less appreciated than the current ones, with the claim "without preservatives" receiving a significantly higher value than the claim "with natural preservatives" ($p < 0.05$), and the claim "with biopreservatives" being assessed as worse than the others ($p < 0.01$) (Figure 2).

The results showed that most participants agreed with the statement about the safety of food biopreservatives. A total of 79% of respondents considered biopreservatives to be less harmful to health than synthetic preservatives, and 55% of them thought that biopreservatives cause less damage to the environment than conventional preservatives. Only 22% of respondents think that biopreservatives improve the flavour of the food compared to the synthetic preservatives. With regards shelf life, only 17% of respondents considered biopreservatives to be better than the synthetic ones. However, the respondents appreciated the use of biopreservatives, especially in the preparation of fruit (71%), fresh-cut vegetables (47%), and meat (42%). They seemed less concerned about processed products. In fact, only 33.6% of respondents said they would prefer the use of biopreservatives in bakery products, and only 6% in processed food products. Finally, the majority of the respondents answered that they were willing to consume biopreservatives only if they replaced the synthetic ones (64.3%). A total of 13.6% thought that they would be willing to consume biopreservatives only if they increased the shelf life of the food products.

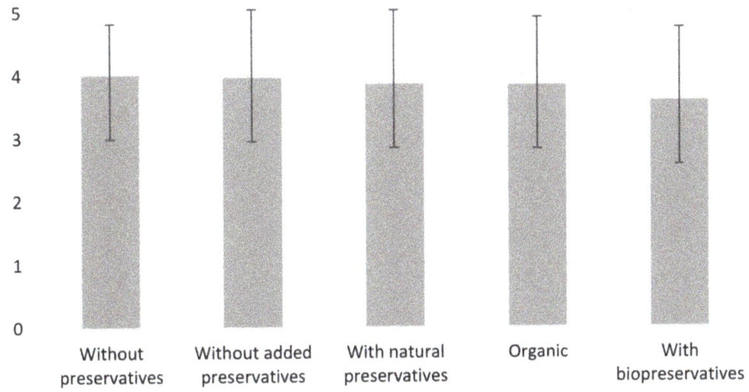

Figure 2. The importance of the claims on hypothetical labels for the product choice (mean and standard deviation). (Attributes importance (scale 1 to 5: where 1 = Not important and 5 = Very important)).

Moreover, we have tried to understand the willingness to pay for food products treated with biopreservatives as preservative replacers. A total of 67.8% are willing to pay for fresh-cut vegetables with "only" biopreservatives; of these, 55.7% are willing to spend between 10–20% more and 12.1% are willing to pay up to 30%. In the case of semi-hard cheeses treated with synthetic preservatives on the rind, 63% of respondents were willing to pay between 10–20% more for the same product treated with biopreservatives, and 13% of respondents were willing to pay up to 30%. Conversely, 24% of respondents were unwilling to pay more for cheese treated with biopreservatives.

To understand the determinants in the preference of a product treated with biopreservatives over conventional products, we have completed our study with a Principal Component Analysis (PCA). This method allowed us to reduce the initial diversity of certain number of variables to a smaller number of Principal Components and to simplify the interpretation of the phenomenon. The PCA can be used when the sample adequacy of the model, indicated by the KMO index [45], is more than 70%; the total variance explained should be greater than 65–70% of the total variance represented by the variables used.

The PCA was applied to variables obtained by the questionnaire answers, representing three main aspects of the phenomenon: sensitivity to label claims (influence of different label claims in the product choice, level of attention paid to different information present on the label and to the price); consumers' behaviour and choices; and personal characteristics. The KMO of our data is equal to 0,703 (Table 1) and the total variance explained reaches a good level (68.35%), with the first eight main components identified (Table 2).

Table 1. The results of the Principal Component Analysis (PCA)—KMO and Bartlett's Test.

Kaiser-Meyer-Olkin Measure of Sampling Adequacy		0.703
Bartlett's Test of Sphericity	Approx. Chi-Square	2638.368
	Df	190
	Sig.	0.000

Table 2. The determinants of consumers' preference towards biopreservatives. A rotated component matrix is shown.

Variables	Components							
	CP1	CP2	CP3	CP4	CP5	CP6	CP7	CP8
Sensitivity to label claims								
Influence of the label: "with natural preservatives"	0.780	0.235	0.165			−0.103		
Influence of the label: "with biopreservatives"	0.766	0.246	0.234			−0.108		
Influence of the label "Organic product"	0.512		0.640	0.148				
Level of attention to Ingredients	0.297	0.739	0.180				−0.142	
Level of attention to the Product origin	0.143	0.570	0.314	0.147		0.207	0.270	0.110
Level of attention to Nutritional Information	0.415	0.545					0.148	0.127
Level of attention to Organic certification	0.378	0.300	0.704					
Level of attention to Brand		0.118	0.104				0.755	
Level of attention to Price			0.140	0.126		0.186	0.722	
Consumers behaviour								
Frequency of reading label		0.819					0.201	0.116
Frequency of organic food consumption		0.190	0.803					0.169
Unwillingness to pay more for a cheese treated with natural preservatives				0.914				
Unwillingness to pay more for a salad treated with natural preservatives			0.106	0.913				0.103
Willingness to pay more for food treated with natural preservatives					0.909	0.116		
Willingness to pay more for food treated with natural preservatives in substitution of synthetic ones	0.115			0.165	0.789	0.269		
Willingness to pay more for food treated with natural preservatives if the shelf life of the product increase						0.842		0.110
Preference for a ready-to-eat fruit salad treated with biopreservatives, expiring at 5 days respect the same product without preservatives expiring at 3 days			0.127		0.225	0.650		0.213
Personal Characteristics								
Age	0.154	0.115	0.460				0.197	0.438
Knowledge about biopreservatives	0.131		0.135					0.679
Gender	0.450					0.122	0.219	0.614

Description of the seven principal main components identified:

CP1 (21.2% of the total variance explained): mainly a female component, susceptible to the indication on the label of "natural preservatives" and "biopreservatives", as well as to organic products and organic certification, where the willingness to pay (WTP) a product treated without synthetic preservatives is weak.

CP2 (8.92% of the total variance explained): a youth component that pays attention to the label, to ingredients, and the origin and nutritional information, but not the brand.

CP3 (7.88% of the total variance explained): older people, with a fair knowledge of the topic, the frequent consumption of organic products, susceptible to the indication on the label of organic products and organic certification and who give little importance to the price.

CP4 (7.48% of the total variance explained): a component with no willingness to pay more for products treated with natural preservatives, no interest in biopreservatives, slightly price sensitive, not interested in organic production or in the products' origin.

CP5 (6.96% of the total variance explained): a component very favorable to the purchase of products treated with biopreservatives, even if they do not replace synthetic preservatives and albeit with a reduced shelf life.

CP6 (5.47% of the total variance explained): a component with high WTP for products treated with natural preservatives if the shelf life of the product increases; generally attentive to the price.

CP7 (5.25% of the total variance explained): a male and youthful component, sensitive to the brand and price and attentive to the origin of the products.

CP8 (5.15% of the total variance explained): a male and elderly component, characterized by a very limited knowledge about biopreservatives, moderately favorable to the use of biopreservatives and having little interest in the origin of the food products.

The factorial scores of the previous eight main components obtained from PCA were used for a Cluster Analysis. A two-step clustering method was applied adopting the squared Euclidean distance algorithm for case processing.

As the eighth component was not very explanatory, we decided to reduce to seven the principal components for the analysis. A four-cluster solution showed the most distinctive profile [46] and was thus the solution retained (Figure 3). The four identified clusters possess an acceptable measure of cohesion and separation, with a silhouette coefficient of 0.43 [47]: Cluster 1 (16.7% of the sample); Cluster 2 (60.3%); Cluster 3 (10.6%); Cluster 4 (12.3%).

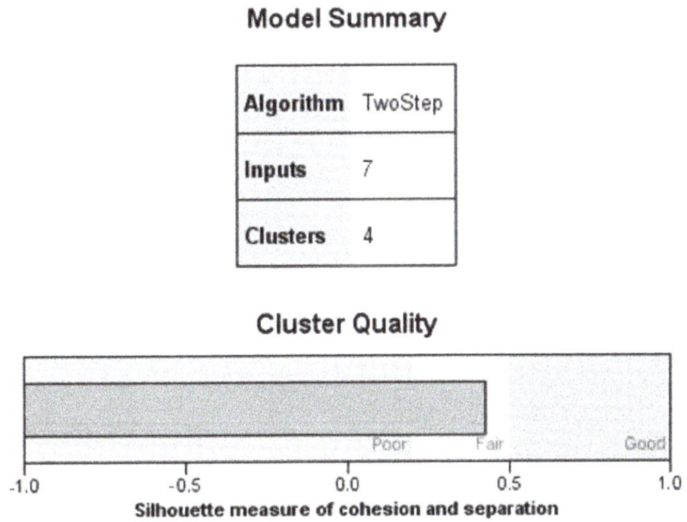

Figure 3. The results of the Cluster Analysis with seven principal components.

Their description is given below (Table 3).

Table 3. Main characteristics of consumers segments (n = 479): mean scores of variables within the groups (standard deviation within brackets).

Clusters	1 (Enthusiastic)		2 (Rational)		3 (Disinterested)		4 (Conscious and Attentive)		Total	
Personal characteristics										
Gender (0 = female, 1= male)	0.45	(0.501)	0.40	(0.491)	0.51	(0.505)	0.37	(0.488)	0.42	(0.494)
Age	38.05	(10.467)	36.75	(11.515)	37.24	(12.239)	35.98	(13.309)	36.923	(11.641)
Knowledge about biopreservatives *	2.35	(0.873)	2.47	(0.858)	2.39	(0.918)	2.56	(1.038)	2.46	(0.890)
Sensitivity to label claims										
Level of attention to Nutritional Information *	3.94	(1.173)	3.72	(1.183)	3.43	(1.118)	3.68	(1.105)	3.72	(1.169)
Level of attention to Brand *	2.64	(1.183)	2.86	(1.120)	2.55	(1.101)	2.86	(1.025)	2.79	(1.120)
Level of attention to Organic certification *	3.63	(1.354)	3.23	(1.249)	2.65	(1.230)	2.81	(1.238)	3.18	(1.292)
Level of attention to Product origin *	4.06	(1.205)	4.00	(1.091)	3.31	(1.516)	3.53	(1.056)	3.88	(1.182)
Level of attention to Price *	3.81	(1.092)	3.59	(1.080)	3.96	(1.183)	4.03	(0.909)	3.72	(1.084)
Level of attention to Ingredients *	4.24	(1.046)	4.16	(1.055)	3.86	(1.281)	3.85	(1.096)	4.10	(1.090)
Consumers behavior										
Frequency of reading label *	3.05	(1.146)	2.94	(1.054)	2.75	(1.214)	2.81	(1.025)	2.92	(1.084)
Frequency of organic food consumption *	1.76	(0.945)	1.72	(0.940)	1.37	(0.958)	1.49	(0.858)	1.66	(0.939)

* Scores on a five-point Likert-type scale (1 = minimum; 5 = maximum).

Enthusiastic (Cluster 1—80 cases): a cluster with less knowledge about the meaning of preservatives and biopreservatives than other individuals, who frequently purchase organic products and declare an unconditional interest in the purchase of products treated with biopreservatives. These consumers are not interested in the shelf life of food products and are willing to pay higher prices than other respondents. It is a niche market with a high prevalence in families of people with professional work, e.g., lawyers, engineers, etc., which would deserve the positioning of a differentiated product of high added value.

Rational (Cluster 2—289 cases): a cluster composed mainly of women (60%), with high educational levels, knowledge of the properties of preservatives, and a high WTP (both for cheese and fresh-cut vegetables treated with biopreservatives). In this group, we find consumers of organic products, attentive to the reading of the label, and in particular to the ingredients and health aspects. The interest in buying is combined with the search for products in which natural preservatives replace synthetic ones.

Disinterested (Cluster 3—51 cases): a cluster with equal representation of men and women, more so than the rest of the sample, with lower-than-average levels of education and very limited knowledge of preservatives. This group expresses a very low WTP, practically zero in the case of cheese and fresh-cut vegetables, consistent with the indication of high importance attributed to the price.

Conscious and attentive (Cluster 4—59 cases): a cluster composed largely of workers or the unemployed, attentive to the price and shelf life of the food products purchased. They show good knowledge about preservatives in food processing, despite a low level of education.

The marketing positioning opportunities of biopreservatives on the consumers basically depends on the strategic objective that the offer (single or aggregate) has. In this context, if the goal is to maximize profit or minimize costs (communication and distribution), the most interesting positions, from the point of view of potential turnover (segment size and WTP), appear to be the clusters 2 and

1. Even cluster 4, although small in size, deserves attention, but it would require a marketing-mix oriented towards products, in particular fruit and vegetables, which are able to combine a good shelf life with low prices.

In general, our results showed that consumers have different levels of understanding of biopreservatives in different situations and do not perceive food additives and biopreservatives in the same way. In fact, as previous literature suggested [4,48,49], consumers hardly recognized food additive information on product labels, showing limited awareness of food additives. The provision of accurate information on the presence/absence of food additives is considered to be an important factor affecting the purchase decision [50,51]. Consumers tend to amplify the risk when a food item or a technology is unknown [52]. Numerous surveys have shown that consumers express concerns about their daily diet, and they are worried about being exposed to synthetic preservatives [43,53]. However, in recent years, consumers have been exploiting new media to become more informed, and social media has become more and more influential in determining their concerns about food quality attributes [54].

Another relevant finding of our study is that the less organic food the respondents consumed, the more they cared for price and the less for the presence of preservatives in convenience foods. Regular organic food consumers preferred biopreservatives compared to synthetic preservatives. However, our results pointed out that even when consumers profess a strong support for environmental attributes, they are still extremely price sensitive. These findings are in line with the literature related to the consumer' approach to the decision to purchase sustainable food [18,55,56].

4. Conclusions

We investigated the factors affecting the acceptance for products treated with EOs as natural preservatives through a direct survey of consumers. Our aim was to analyse consumer knowledge and perception of the information currently used by companies and to verify acceptance of biopreservatives. Moreover, consumer acceptance and willingness to pay were analysed with respect to shelf life and replacement of synthetic preservatives with natural preservatives.

The results of our direct survey highlighted the difficulty in perceiving the exact meaning of the term "biopreservative", which, however, was generally associated with positive opinions such as the reduction of damage to health or positive environmental impact. Moreover, the term had a lower impact on the consumer than the claim "without preservatives".

Our results suffer from two main limitations. Firstly, the sample analyzed in this study is not representative of the whole Italian population. However, the relationships between the variables analyzed and the consumers' perceptions about biopreservatives remain valid and allow us to obtain interesting results. Secondly, there is not a clear understanding by consumers of biopreservatives and our study might suffer from hypothetical bias which could have affected the estimation of consumers' acceptance [10,41,42].

The findings of this research constitute an opportunity for food companies; suppliers of foods added with biopreservatives should thus concentrate on organic and high-quality foods with a low level of processing. The results obtained could also help scientists in addressing the research in this field, with the aim of meeting the requirements of both consumers and food industries. Moreover, this study also expresses the need for a consumer campaign and a better education on food biopreservatives.

This study shows how research on consumer preferences and priorities (e.g., naturalness vs. shelf life; biopreservatives vs. price; etc.) can be of paramount importance to promote the scientific and technological evolution of food manufacturing toward practical applications of biopreservatives, which could be fundamental in the future market for their potential to decrease the negative impact of foods on health and environment.

Author Contributions: Conceptualization, M.A.P., E.C. and A.F.; Data curation, M.A.P., E.C. and A.F.; Funding acquisition, A.P.; Investigation, M.A.P.; Methodology, M.A.P. and A.F.; Software, A.F.; Supervision, M.A.P., A.P.;

Writing—original draft, M.A.P.; Writing—review & editing, M.A.P., E.C. and A.S. All authors have read and agreed to the published version of the manuscript.

Funding: This research was supported by the grant of Regione Abruzzo, for the project "PSR 2017–2013 Misura 1.2.4" cod. CUA 2446850691, title "Sviluppo di sistemi convenzionali e innovazioni per la produzione di composti bioattivi da materie prime vegetali per l'impiego nel settore alimentare".

Conflicts of Interest: The authors declare no conflict of interest.

: **Data Availability:** The data supporting the findings of this study are available from the corresponding author (M.A.P.) and the first author (M.A.P.) upon reasonable request.

References

1. Regulation (EC) No. 1333/2008 of the European Parliament and of the Council of 16 December 2008 on Food Additives. Available online: https://eur-lex.europa.eu/legal-content/IT/TXT/HTML/?uri=CELEX:32008R1333&from=EN (accessed on 5 May 2020).
2. Additives in Dairy Foods; Consumer Perceptions of Additives in Dairy Products. In *Encyclopedia of Dairy Sciences*, 2nd ed.; Academic Press: Cambridge, MA, USA, 2011. [CrossRef]
3. Arbindra, R.; Stanley, M.; Fletcher, K.H.; McWatters, S.; Misra, K.; Deodhar, E. Perception of food safety and change in food consumption habits: A consumer analysis. *Int. J. Consum.* **2001**, *25*, 43–52. [CrossRef]
4. Shim, S.M.; Seo, S.H.; Lee, Y.; Moon, G.I.; Kim, M.S.; Park, J.H. Consumers' knowledge and safety perceptions of food additives: Evaluation on the effectiveness of transmitting information on preservatives. *Food Control* **2011**, *22*, 1054–1060. [CrossRef]
5. Devcich, D.A.; Pedersen, I.K.; Petrie, K.J. You eat what you are: Modern health worries and the acceptance of natural and synthetic additives in functional foods. *Appetite* **2007**, *48*, 333–337. [CrossRef] [PubMed]
6. Carocho, M.; Barreiro, M.F.; Morales, P.; Ferreira, I.C.F.R. Adding molecules to Food, pros and cons: A review of synthetic and natural food additives. *Compr. Rev. Food Sci. Food Saf.* **2014**, *13*, 377–399. [CrossRef]
7. Bearth, A.; Cousin, M.; Siegrist, M. The consumer's perception of artificial food additives: Influences on acceptance, risk and benefit perceptions. *Food Qual. Prefer.* **2014**, *38*, 14–23. [CrossRef]
8. Carocho, M.; Morales, P.; Ferreira, I.C. Natural food additives: Quo vadis? *Trends Food Sci. Technol.* **2015**, *45*, 284–295. [CrossRef]
9. Pellegrini, M.; Ricci, A.; Serio, A.; Chaves López, C.; Mazzarrino, G.; D'Amato, S.; Lo Sterzo, C.; Paparella, A. Characterization of essential oils obtained from Abruzzo autochthonous plants: Antioxidant and antimicrobial activities assessment for food application. *Foods* **2018**, *7*, 19. [CrossRef]
10. Perito, M.A.; Di Fonzo, A.; Sansone, M.; Russo, C. Consumer acceptance of food obtained from olive by-products. *Brit. Food J.* **2019**, *122*, 212–226. [CrossRef]
11. Coderoni, S.; Perito, M.A. Sustainable consumption in the circular economy. An analysis of consumers' purchase intentions for waste-to-value food. *J. Clean. Prod.* **2020**, *252*, 119870. [CrossRef]
12. European Food Safety Authority (EFSA). Use of rosemary extracts as a food additive-Scientific Opinion of the Panel on Food Additives, Flavourings, Processing Aids and Materials in Contact with Food. *EFSA J.* **2008**, *6*, 1–29. [CrossRef]
13. Lockie, S.; Lyons, K.; Lawrence, G.; Grice, J. Choosing organics: A path analysis of factors underlying the selection of organic food among Australian consumers. *Appetite* **2004**, *43*, 135–146. [CrossRef] [PubMed]
14. Rozin, P. The meaning of "natural": Process more important than content. *Psychol. Sci.* **2005**, *16*, 652–658. [CrossRef] [PubMed]
15. Siipi, H. Is natural food healthy? *J. Agric. Environ. Ethics* **2013**, *26*, 797–812. [CrossRef]
16. Palmieri, N.; Perito, M.A. Consumers' willingness to consume sustainable and local wine in Italy. *Ital. J. Food Sci.* **2020**, *32*, 222–233. [CrossRef]
17. Janssen, M. Determinants of organic food purchases: Evidence from household panel data. *Food Qual. Prefer.* **2018**, *68*, 19–28. [CrossRef]
18. Defrancesco, E.; Perito, M.A.; Bozzolan, I.; Cei, L.; Stefani, G. Testing consumers' preferences for environmental attributes of pasta. Insights from an ABR approach. *Sustainability* **2017**, *9*, 1701. [CrossRef]
19. Willer, H.; Lernoud, J. *The World of Organic Agriculture. Statistics and Emerging Trends 2017*; Research Institute of Organic Agriculture (FiBL), Frick, and IFOAM—Organics International: Bonn, Germany, 2017.

20. Jung, Y.J.; Padmanabahn, A.; Hong, J.H.; Lim, J.; Kim, K.O. Consumer freshness perception of spinach samples exposed to different storage conditions. *Postharvest Biol. Technol.* **2012**, *73*, 115–121. [CrossRef]
21. Bertazzoli, A.; Buccioli, N.; Nocella, G. Consumers perception and evaluation of fresh cut buying attributes: A survey on the Italian market. *J. Food Prod. Market* **2005**, *11*, 35–55. [CrossRef]
22. D'Amato, S.; Serio, A.; Chaves López, C.; Paparella, A. Hydrosols: Biological activity and potential as antimicrobials for food applications. *Food Control* **2018**, *86*, 126–137. [CrossRef]
23. Boutouis, M.Z.; Hammoudi, A.; Benhassine, W.; Perito, M.A. Uncertainty of food contamination origin and liability rules: Implications for bargaining power. *Agribusiness* **2018**, *34*, 77–92. [CrossRef]
24. Rossi, C.; Chaves López, C.; Smole Mozina, S.; Di Mattia, C.; Scuota, S.; Luzzi, I.; Jenic, T.; Paparella, A.; Serio, A. *Salmonella enterica* adhesion: Effect of *Cinnamomum zeylanicum* essential oil on lettuce. *LWT* **2019**, *111*, 16–22. [CrossRef]
25. Mazzarrino, G.; Paparella, A.; Chaves López, C.; Faberi, A.; Sergi, M.; Sigismondi, C.; Compagnone, D.; Serio, A. *Salmonella enterica* and *Listeria monocytogenes* inactivation dynamics after treatment with selected essential oils. *Food Control* **2015**, *50*, 794–803. [CrossRef]
26. Grande-Tovar, C.D.; Serio, A.; Delgado_Ospina, J.; Paparella, A.; Rossi, C.; Chaves López, C. Chitosan films incorporated with *Thymus capitatus* essential oil: Mechanical properties and antimicrobial activity against degradative bacterial species isolated from tuna (*Thunnus* sp.) and swordfish (*Xiphias gladius*). *J. Food Sci. Technol.* **2018**, *55*, 4256–4265. [CrossRef] [PubMed]
27. Hyldgaard, M.; Mygind, T.; Meyer, R.L. Essential oils in food preservation: Mode of action, synergies, and interactions with food matrix components. *Front. Microbiol.* **2012**, *3*, 12. [CrossRef]
28. Nait Mohand, N.; Hammoudi, A.; Radjef, M.S.; Hamza, O.; Perito, M.A. How do food safety regulations influence market price? A theoretical analysis. *Brit. Food J.* **2017**, *119*, 1687–1704. [CrossRef]
29. Regulation (EC) No 1334/2008 of the European Parliament and of the Council of 16 December 2008 on Flavourings and Certain Food Ingredients with Flavouring Properties for Use in and on Foods and Amending Council Regulation (EEC) No 1601/91, Regulations (EC) No 2232/96 and (EC) No 110/2008 and Directive 2000/13/EC. Available online: https://eur-lex.europa.eu/legal-content/EN/ALL/?uri=CELEX%3A32008R1334 (accessed on 5 May 2020).
30. Regulation (EU) No 1169/2011 of the European Parliament and of the Council of 25 October 2011 on the Provision of Food Information to Consumers. Available online: https://eur-lex.europa.eu/legal-content/EN/TXT/PDF/?uri=CELEX:32011R1169 (accessed on 5 May 2020).
31. Van Boxstael, S.; Devlieghere, F.; Berkvens, D.; Vermeulen, A.; Uyttendaele, M. Understanding and attitude regarding the shelf life labels and dates on pre-packed food products by Belgian consumers. *Food Control* **2014**, *37*, 85–92. [CrossRef]
32. Newsome, R.; Balestrini, C.G.; Baum, M.D.; Corby, J.; Fisher, W.; Goodburn, K.; Labuza, T.P.; Prince, G.; Thesmar, H.S.; Yiannas, F. Applications and perceptions of date labeling of food. *Compr. Rev. Food Sci. Food Saf.* **2014**, *13*, 745–769. [CrossRef]
33. Stranieri, S.; Baldi, L. Shelf life date extension of fresh-cut salad: A consumer perspective. *J. Food Prod. Mark.* **2017**, *23*, 939–954. [CrossRef]
34. Vidal, L.; Ares, G.; Giménez, A. Projective techniques to uncover consumer perception: Application of three methodologies to ready-to-eat salads. *Food Qual. Prefer.* **2013**, *28*, 1–7. [CrossRef]
35. Dinnella, C.; Torri, L.; Caporale, G.; Monteleone, E. An exploratory study of sensory attributes and consumer traits underlying liking for and perceptions of freshness for ready to eat mixed salad leaves in Italy. *Food Res. Int.* **2014**, *59*, 108–116. [CrossRef]
36. García-Díez, J.; Alheiro, J.; Pinto, A.L.; Soares, L.; Falco, V.; Fraqueza, M.J.; Patarata, L. The Impact of essential oils on consumer acceptance of Chouriço de vinho–a dry-cured sausage made from wine-marinated meat–assessed by the Hedonic Scale, JAR intensity scale and consumers' "Will to Consume and Purchase". *J. Food Process Pres.* **2017**, *41*, e13056. [CrossRef]
37. Van Gunst, A.; Roodenburg, A.J. Consumer Distrust about E-numbers: A Qualitative Study among Food Experts. *Foods* **2019**, *8*, 178. [CrossRef] [PubMed]
38. Albertsen, L.; Wiedmann, K.P.; Schmidt, S. The Impact of Innovation-related Perception on Consumer Acceptance of Food Innovations-Development of an Integrated Framework of the Consumer Acceptance Process. *Food Qual. Prefer.* **2020**, *84*, 103958. [CrossRef]

39. Chiodo, E.; Casolani, N.; Fantini, A. Regulatory policies and consumers quality perception in the wine sector. *Enometrica* **2011**, *4*, 39–57.
40. Casolani, N.; Greehy, G.M.; Fantini, A.; Chiodo, E.; McCarthy, M.B. Consumer perceptions of nanotechnology applications in Italian wine. *Ital. J. Food Sci.* **2015**, *27*, 93–107. [CrossRef]
41. Palmieri, N.; Perito, M.A.; Macrì, M.C.; Lupi, C. Exploring consumers' willingness to eat insects in Italy. *Brit. Food J.* **2019**, *121*, 2937–2950. [CrossRef]
42. Palmieri, N.; Perito, M.A.; Lupi, C. Consumer acceptance of cultured meat: Some hints from Italy. *Brit. Food J.* **2020**. [CrossRef]
43. Dickson-Spillmann, M.; Siegrist, M.; Keller, C. Attitudes toward chemicals are associated with preference for natural food. *Food Qual. Prefer.* **2011**, *22*, 149–156. [CrossRef]
44. Palmieri, N.; Forleo, M.B. The potential of edible seaweed within the western diet. A segmentation of Italian consumers. *Int. J. Gastron. Food Sci.* **2020**, *20*, 100202. [CrossRef]
45. Kaiser, H.F. An index of factorial simplicity. *Psychometrika* **1974**, *39*, 31–36. [CrossRef]
46. Hair, J.F.; Black, W.C.; Babin, B.J.; Anderson, R.E.; Tatham, R.L. *Multivariate Data Analysis*, 5th ed.; Prentice Hall: Upper Saddle River, NJ, USA, 1998; pp. 207–219.
47. Rousseeuw, P.J. Silhouettes: A Graphical Aid to the Interpretation and Validation of Cluster Analysis. *J. Comput. Appl. Math.* **1998**, *20*, 53–65. [CrossRef]
48. Song, H.; Schwarz, N. If it's difficult to pronounce, it must be risky: Fluency, familiarity, and risk perception. *Psychol. Sci.* **2009**, *20*, 135–138. [CrossRef] [PubMed]
49. Kaptan, B.; Kayısoglu, S. Consumers' attitude towards food additives. *Am. J. Food Sci. Nutr. Res.* **2015**, *2*, 21–25.
50. Kim, E.J.; Na, H.J.; Kim, Y.N. Awareness on food additives and purchase of processed foods containing food additives in middle school students. *Korean J. Hum. Ecol.* **2007**, *16*, 205–218. [CrossRef]
51. Aoki, K.; Shen, J.; Saijo, T. Consumer reaction to information on food additives: Evidence from an eating experiment and a field survey. *Econ. Behav. Organ* **2010**, *73*, 433–438. [CrossRef]
52. Grunert, K.G. Food quality and safety: Consumer perception and demand. *Eur. Rev. Agric. Econ.* **2005**, *32*, 369–391. [CrossRef]
53. European Commission. Special Eurobarometer: Health and Food, 2006. Retrieved 19 September 2008. Available online: http://ec.europa.eu/health/ph_publication/eb_food_en.pdf (accessed on 8 February 2020).
54. Russo, C.; Simeone, M.; Perito, M.A. Educated Millennials and Credence Attributes of Food Products with Genetically Modified Organisms: Knowledge, Trust and Social Media. *Sustainability* **2020**, *12*, 8534. [CrossRef]
55. Do Paco, A.; Raposo, M. "Green" segmentation: An application to the Portuguese consumer market. *Mark. Intell. Plann.* **2009**, *27*, 364–379. [CrossRef]
56. Perito, M.A.; Coderoni, S.; Russo, C. Consumer Attitudes towards Local and Organic Food with Upcycled Ingredients: An Italian Case Study for Olive Leaves. *Foods* **2020**, *9*, 1325. [CrossRef]

Publisher's Note: MDPI stays neutral with regard to jurisdictional claims in published maps and institutional affiliations.

MDPI
St. Alban-Anlage 66
4052 Basel
Switzerland
Tel. +41 61 683 77 34
Fax +41 61 302 89 18
www.mdpi.com

Sustainability Editorial Office
E-mail: sustainability@mdpi.com
www.mdpi.com/journal/sustainability

www.ingramcontent.com/pod-product-compliance
Lightning Source LLC
Chambersburg PA
CBHW062057220926
51850CB00010B/2794
* 9 7 8 3 0 3 6 5 6 0 1 3 7 *